Plant Hormones

METHODS IN MOLECULAR BIOLOGY™

John M. Walker, SERIES EDITOR

502. Bacteriophages: *Methods and Protocols, Volume 2: Molecular and Applied Aspects,* edited by *Martha R. J. Clokie and Andrew M. Kropinski 2009*
501. Bacteriophages: *Methods and Protocols, Volume 1: Isolation,* Characterization, and Interactions, edited by *Martha R. J. Clokie and Andrew M. Kropinski 2009*
496. DNA and RNA Profiling in Human Blood: *Methods and Protocols,* edited by *Peter Bugert, 2009*
495. Plant Hormones: *Methods and Protocols, Second Edition,* edited by *Sean Cutler and Dario Bonetta, 2009*
493. Auditory and Vestibular Research: *Methods and Protocols,* edited by *Bernd Sokolowski, 2009*
490. Protein Structure, Stability, and Interactions, edited by *John W. Shriver, 2009*
489. Dynamic Brain Imaging: *Methods and Protocols,* edited by *Fahmeed Hyder, 2009*
485. HIV Protocols: *Methods and Protocols,* edited by *Vinayaka R. Prasad and Ganjam V. Kalpana, 2009*
484. Functional Proteomics: *Methods and Protocols,* edited by *Julie D. Thompson, Christine Schaeffer-Reiss, and Marius Ueffing, 2008*
483. Recombinant Proteins From Plants: *Methods and Protocols,* edited by *Lóic Faye and Veronique Gomord, 2008*
482. Stem Cells in Regenerative Medicine: *Methods and Protocols,* edited by *Julie Audet and William L. Stanford, 2008*
481. Hepatocyte Transplantation: *Methods and Protocols,* edited by *Anil Dhawan and Robin D. Hughes, 2008*
480. Macromolecular Drug Delivery: *Methods and Protocols,* edited by *Mattias Belting, 2008*
479. Plant Signal Transduction: *Methods and Protocols,* edited by *Thomas Pfannschmidt, 2008*
478. Transgenic Wheat, Barley and Oats: *Production and Characterization Protocols,* edited by *Huw D. Jones and Peter R. Shewry, 2008*
477. Advanced Protocols in Oxidative Stress I, edited by *Donald Armstrong, 2008*
476. Redox-Mediated Signal Transduction: *Methods and Protocols,* edited by *John T. Hancock, 2008*
475. Cell Fusion: *Overviews and Methods,* edited by *Elizabeth H. Chen, 2008*
474. Nanostructure Design: *Methods and Protocols,* edited by *Ehud Gazit and Ruth Nussinov, 2008*
473. Clinical Epidemiology: *Practice and Methods,* edited by *Patrick Parfrey and Brendon Barrett, 2008*
472. Cancer Epidemiology, Volume 2: *Modifiable Factors,* edited by *Mukesh Verma, 2008*
471. Cancer Epidemiology, Volume 1: *Host Susceptibility Factors,* edited by *Mukesh Verma, 2008*
470. Host-Pathogen Interactions: *Methods and Protocols,* edited by *Steffen Rupp and Kai Sohn, 2008*
469. Wnt Signaling, Volume 2: *Pathway Models,* edited by *Elizabeth Vincan, 2008*
468. Wnt Signaling, Volume 1: *Pathway Methods and Mammalian Models,* edited by *Elizabeth Vincan, 2008*
467. Angiogenesis Protocols: *Second Edition,* edited by *Stewart Martin and Cliff Murray, 2008*
466. Kidney Research: *Experimental Protocols,* edited by *Tim D. Hewitson and Gavin J. Becker, 2008*
465. Mycobacteria, Second Edition, edited by *Tanya Parish and Amanda Claire Brown, 2008*
464. The Nucleus, Volume 2: *Physical Properties and Imaging Methods,* edited by *Ronald Hancock, 2008*
463. The Nucleus, Volume 1: *Nuclei and Subnuclear Components,* edited by *Ronald Hancock, 2008*
462. Lipid Signaling Protocols, edited by *Banafshe Larijani, Rudiger Woscholski, and Colin A. Rosser, 2008*
461. Molecular Embryology: *Methods and Protocols, Second Edition,* edited by *Paul Sharpe and Ivor Mason, 2008*
460. Essential Concepts in Toxicogenomics, edited by *Donna L. Mendrick and William B. Mattes, 2008*
459. Prion Protein Protocols, edited by *Andrew F. Hill, 2008*
458. Artificial Neural Networks: *Methods and Applications,* edited by *David S. Livingstone, 2008*
457. Membrane Trafficking, edited by *Ales Vancura, 2008*
456. Adipose Tissue Protocols, Second Edition, edited by *Kaiping Yang, 2008*
455. Osteoporosis, edited by *Jennifer J. Westendorf, 2008*
454. SARS- and Other Coronaviruses: *Laboratory Protocols,* edited by *Dave Cavanagh, 2008*
453. Bioinformatics, Volume 2: *Structure, Function, and Applications,* edited by *Jonathan M. Keith, 2008*
452. Bioinformatics, Volume 1: *Data, Sequence Analysis, and Evolution,* edited by *Jonathan M. Keith, 2008*
451. Plant Virology Protocols: *From Viral Sequence to Protein Function,* edited by *Gary Foster, Elisabeth Johansen, Yiguo Hong, and Peter Nagy, 2008*
450. Germline Stem Cells, edited by *Steven X. Hou and Shree Ram Singh, 2008*
449. Mesenchymal Stem Cells: *Methods and Protocols,* edited by *Darwin J. Prockop, Douglas G. Phinney, and Bruce A. Brunnell, 2008*
448. Pharmacogenomics in Drug Discovery and Development, edited by *Qing Yan, 2008*
447. Alcohol: *Methods and Protocols,* edited by *Laura E. Nagy, 2008*
446. Post-translational Modifications of Proteins: *Tools for Functional Proteomics, Second Edition,* edited by *Christoph Kannicht, 2008*
445. Autophagosome and Phagosome, edited by *Vojo Deretic, 2008*
444. Prenatal Diagnosis, edited by *Sinhue Hahn and Laird G. Jackson, 2008*
443. Molecular Modeling of Proteins, edited by *Andreas Kukol, 2008*

METHODS IN MOLECULAR BIOLOGY™

Plant Hormones

Methods and Protocols, Second Edition

Edited by

Sean Cutler

Department of Botany and Plant Sciences,
University of California at Riverside, Riverside, USA

and

Dario Bonetta

Faculty of Science, University of Ontario Institute of Technology,
Oshawa, ON, Canada

Humana Press

Editors
Sean Cutler
Department of Botany and Plant Sciences
University of California at Riverside
Riverside, USA

Dario Bonetta
Faculty of Science
University of Ontario Institute of Technology
Oshawa, ON, Canada

Series Editor
John M. Walker
School of Life Sciences
University of Hertfordshire
Hatfield, Hertfordshire, AL10 9AB, UK

ISSN 1064-3745
e-ISSN 1940-6029
ISBN 978-1-61737-930-7
e-ISBN 978-1-59745-477-3
DOI 10.1007/978-1-59745-477-3

Printed on acid-free paper

springer.com

Preface

The last 10 years have witnessed an explosion in our understanding of plant hormones. The often vague models of hormone action developed over decades have been replaced in short order by detailed molecular models that include receptors and in many cases downstream signal transduction components. Given the rapid progress in understanding the mechanism of action of plant growth regulators, a technical review of hormone methodology is timely. Our book focuses on genetic, biochemical, analytical and chemical biological approaches for understanding and dissecting plant hormone action. The greatest strides in plant hormone biology have come, by and large, from the use of genetic methods to identify receptors and we dedicate a chapter to general genetic methods of analysis using the model system Arabidopsis thaliana. A cluster of chapters focuses on biochemical methods for documenting interactions between hormones and their receptors. The importance of these assays is tremendous; receptor–ligand interactions in animal model systems have been the cornerstones of pharmacological and medicinal chemical assays that have enabled identification of selective and non-selective agonists and antagonists that can be used to further probe and dissect questions of receptor function. This is likely to be a major new frontier in plant hormone research. Given these recent assays for plant hormone receptors, the time is ripe to investigate chemical biological methods for exploiting these assays to develop an understanding the mechanism of action of synthetic plant growth regulators; in this context, we offer methods for conducting chemical genetic screens to identify new growth regulators in Arabidopsis and other plant species. Lastly, the last few years have also seen impressive growth in analytical methods for measuring plant hormones; we offer two chapters on both LC-MS and GC-MS based methods for measuring endogenous plant hormone levels. Collectively, these protocols should empower plant hormone biologists to dissect the ext generation of questions in plant hormone biology and signal transduction.

S. Cutler
D. Bonetta

Contents

Contributors

SUZANNE R. ABRAMS • *National Research Council of Canada – Plant Biotechnology Institute, Saskatoon, Saskatchewan, Canada.*
GEORGE W. BASSEL • *Department of Cell & Systems Biology, University of Toronto, Toronto, ON, Canada.*
THOMAS BERLETH • *Department of Cell & Systems Biology, University of Toronto, Toronto, ON, Canada.*
ANA CAÑO-DELGADO • *Departamento de Genética Molecular. Consorcio CSIC-IRTA, Barcelona, Spain.*
PADMANABHAN CHELLAPPAN • *Department of Plant Pathology and Microbiology, Center for Plant Cell Biology and Institute for Integrative Genome Biology, University of California Riverside, CA, USA.*
ATSUSHI HANADA • *Growth Regulation Research Group, RIKEN Plant Science Center, Tsurumi, Yokohama, Japan.*
GLENN R. HICKS • *Center for Plant Cell Biology and Department of Botany and Plant Sciences at University of California Riverside. CA, USA.*
MASAYUKI HIGUCHI • *Laboratory of Plant Molecular Genetics, Nara Institute of Science and Technology, Takayama, Ikoma, Japan.*
ROBERT D. HILL • *Plant Science Department , University of Manitoba, Winnipeg, MB, Canada.*
HAILING JIN • *Department of Plant Pathology and Microbiology, Center for Plant Cell Biology and Institute for Integrative Genome Biology, University of California Riverside, CA, USA.*
TATSUO KAKIMOTO • *Department of Biological Sciences, Graduate School of Science, Osaka University, Toyonaka, Osaka, Japan.*
YUJI KAMIYA • *Growth Regulation Research Group, RIKEN Plant Science Center, Tsurumi, Yokohama, Japan.*
STEFAN KEPINSKI • *Centre for Plant Sciences, University of Leeds, Leeds, LS2 9JT, UK*
SERGEY N. LOMIN • *Institute of Plant Physiology, Russian Academy of Sciences, Moscow, Russia.*
DANIELLE MARCOS • *Department of Cell & Systems Biology, University of Toronto, Toronto, ON, Canada.*
PETER MCCOURT • *Department of Cell & Systems Biology, University of Toronto, Toronto, ON. Canada.*
TAKESHI MIZUNO • *Laboratory of Molecular Microbiology, School of Agriculture, Nagoya University, Chikusa-ku, Nagoya, Japan.*
EIJI NAMBARA • *Department of Cell and Systems Biology, University of Toronto, Toronto ON, Canada*
LORENA NORAMBUENA • *Center for Plant Cell Biology and Department of Botany and Plant Sciences at University of California Riverside. CA, USA.*

JULIAN NORTHEY • *Department of Cell & Systems Biology, University of Toronto, Toronto, ON. Canada.*
MASANORI OKAMOTO • *Growth Regulation Research Group, RIKEN Plant Science Center, Tsurumi, Yokohama, Japan.*
STACEY J. OWEN • *National Research Council of Canada – Plant Biotechnology Institute, Saskatoon, Saskatchewan, Canada.*
NICHOLAS J. PROVART • *Department of Cell & Systems Biology, University of Toronto, Toronto, ON, Canada.*
NATASHA V. RAIKHEL • *Center for Plant Cell Biology and Department of Botany and Plant Sciences at University of California Riverside. CA, USA.*
FAWZI A. RAZEM • *Plant Science Department, University of Manitoba, Winnipeg, MB, Canada.*
GEORGY A. ROMANOV • *Institute of Plant Physiology, Russian Academy of Sciences, Moscow, Russia.*
ZHI-YONG WANG • *Department of Plant Biology, Carnegie Institution of Washington, Stanford, CA, USA.*
SHINJIRO YAMAGUCHI • *Growth Regulation Research Group, RIKEN Plant Science Center, surumi, Yokohama, Japan.*

Chapter 1

Using Reverse Genetics to Develop Small Knockout Collections for Specific Biological Questions

Julian Northey and Peter McCourt

Abstract

With the advent of indexed mutagenized insertion lines in Arabidopsis, it is now possible to order small knockout collections of particular genes to probe a question of biological interest. This first requires querying Arabidopsis databases to identify lines of interest, ordering them and then verifying homozygous lines to make your collection. Once the collection is in hand, it can be used multiple times to test scientific hypotheses as they arise.

Key words: Arabidopsis, knockout, T-DNA insertion, genotyping, SALK, SIGnAL, homozygous, reverse genetics, gene family.

1. Introduction

In the early 1980s, *Arabidopsis thaliana* (Arabidopsis) became the model organism of choice for many plant biologists because of the ability to do simple genetic screens. These screens were based on forward genetics in which mutagenized seeds or plants were assayed for developmental, biochemical or physiological defects and then the subsequent mutant alleles were molecularly identified; usually by tedious chromosome walking. The complete sequencing of the *Arabidopsis* genome in 2000, in combination with a facile transformation system, changed the paradigm of genetics screens in *Arabidopsis* because it was possible to identify where a T-DNA from a transformation event had inserted itself into the *Arabidopsis* genome. Since each T-DNA insertion was a molecularly tagged mutagenic event, a collection of gene knockout lines could be generated in which the insertion site of the T-DNA could be easily determined.

Sean Cutler, Dario Bonetta (eds.), *Plant Hormones: Methods and Protocols, Second Edition, vol. 495*

DOI 10.1007/978-1-59745-477-3_1 Springerprotocols.com

Unlike forward genetics screens that seek to elucidate the sequence of a mutant gene from an altered phenotype, these T-DNA lines ushered in the era of reverse genetics in which the insertion event was first molecularly identified and then the line was scored for potential phenotypes. The process of developing and utilizing these mutagenic insertional approaches resulted in the systematic creation of large collections of gene-indexed 'knockout' mutations.

With the increasing efficiency and efficacy of a reverse genetics approach, systematic studies of gene function for a given gene family are becoming more feasible. For example, by systematically identifying knockout mutations in cytokinin response, combinations of triple and quadruple mutants were constructed through genetic crosses to determine the functions of the family *(1)*. Although this type of approach has become commonplace in the literature, researchers are now expanding their criteria for classifications. In a recent study on stomatal development, for example, transcriptome analysis was performed in two mutant backgrounds to first identify potential stomatal regulatory genes that were differentially regulated *(2)*. From this list, two alleles (160 T-DNA insertions representing 82 genes) were ordered and directly screened for stomatal defects and ten lines were identified. This case exemplified screening the list of insertion lines for potentially interesting phenotypes before verifying that the line has the correct insertion. A third approach that has been funded extensively by the *Arabidopsis* 2010 project in recent years has involved taking all the genes involved in a specific biochemical process, such as kinases for example, and creating a knockout collection of these for future phenotyping *(3)*. In all these cases, instead of identifying a particular knockout in a specific gene, lists of genes are identified that can range from a small family in the single digits to lists that represent hundreds of genes. Although the size of the gene knockout list can vary, we are now entering a phase of *Arabidopsis* reverse genetics where optimal approaches can be explored to generate specific knockout collections tailored to a researcher's question. In this chapter, we explore ways of generating small-targeted knockout collections of 50–100 genes.

Generally two contrasting approaches can be considered when generating a knockout collection of more than 50 genes. As mentioned in the stomatal screen, the first approach was to order two alleles for each gene directly form the stock center and screen them for an expected genotype without confirming their genotypes. If a phenotype is found in both alleles in these mixed populations, then these specific lines are advanced and characterized. The advantage of this approach is that it is less labor intensive, but it also means that subtle phenotypes will most likely be missed. By contrast, it is possible to first screen homozygous mutants for a given gene, followed by a phenotypic screen of choice. One would then have a 'knockout' collection that could be used for other screens. However, verification of homozygous lines for a large number of genes can be laborious and time consuming. In either case, the inevitable outcome of a reverse genetic screen using any of the aforementioned collections, there must be a clear link between genotype and phenotype; that is, one must prove that the phenotype being characterized is indeed controlled by the gene of interest.

2. Materials

1. Primers: Store stock and working solutions at –20°C.

   Lba1 5' –TGG TTC ACG TAG TGG GCC ATC G- 3'

   Lbb1 5' –GCG TGG ACC GCT TGC TGC AAC T- 3'

   Lbc1 5' –GAA CAA CAC TCA ACC CTA TCT CGG- 3'
2. PCR tubes (ABgene).
3. Agarose.
4. Kanamycin (Sigma).
5. Isopropanol (Sigma).
6. Chloroform (Sigma).
7. Ethanol.
8. Liquid nitrogen.
9. β-mercaptoethanol (Sigma).
10. Plastic microtube pestles (Ultident).
11. Dremel.
12. CTAB buffer (per 200 ml): 4 g CTAB, 16.4 g NaCl, 8 ml 0.5 M EDTA, 20 ml 1 M Tris–HCl (pH 8.0). Autoclave, store at room temperature, and keep sterile between each use.
13. Extraction buffer (per 200 ml): 40 ml 1 M Tris–HCl (pH 8.0), 10 ml 5 M NaCl, 10 ml 0.5 M EDTA, 5 ml 20% SDS, 135 ml sterile H_2O.

3. Methods

T-DNA insertional mutagenesis has a number of advantages and disadvantages that need to be considered when it comes to organizing a reverse genetic screen for large gene families and/or genes clustered based on a particular biological process (e.g., >50) *(4, 5)*. Insertion of T-DNA into the genome is usually accompanied by a functional selectable marker (i.e., kanamycin), which can be used for initial screening (approximately 25% show no kanamycin resistance – 800 lines tested; personal data) (*see* **Note 1**). Another advantage is that the T-DNA fragment can be used as a molecular tag for the insertion site, thereby allowing PCR-based screening to look for homozygous mutants. A considerable disadvantage to T-DNA insertional mutagenesis is the fact that there is an average of 1.5 insertions per T-DNA mutant, and often there are complex integration patterns of T-DNA. This is especially relevant if a phenotype of interest is found and only one homozygous line can be isolated for that given gene. In this case, considerable care must be taken to establish the link between genotype and phenotype (*see* **Section 3.5**).

One approach is to begin by screening for homozygous SALK T-DNA lines. Appropriate T-DNA lines must be chosen, primers designed for characterization of T-DNA mutants, followed by genomic DNA extraction and PCR-based screening for homozygotes.

3.1. Selecting and Ordering SALK T-DNA Lines

Given that there is a growing collection of SALK homozygous T-DNA lines being deposited each month, it is first advisable to look through their collection for any lines that you may be interested in (methylome.salk.edu/cgi-bin/homozygotes.cgi). For your remaining AGIs (i.e., those without a homozygous T-DNA insertion), visit the "SIGnAL 'T-DNA Express' *Arabidopsis* Gene Mapping Tool" webpage (signal.salk.edu/cgi-bin/tdnaexpress). Here, you can enter your AGIs into the relevant search box entitled 'Gene Name'. After performing your search, above this search box there will be a graphical display outlining all the available resources for that given gene. Three different T-DNA collections are shown here; SALK, SAIL, and GABI. SALK and SAIL lines can be ordered through the TAIR website (you will need an account at TAIR to do so). GABI lines must be ordered through the GABI website with an account (www.gabi-kat.de/db/seed_request.php). Alternatively, for a large list of genes, one can do a multiple-gene search at: signal.salk.edu/tdnamsearch.html. The output of data from this search can then be sorted and organized in spreadsheet for easier selection of the 'best' T-DNA lines available. If possible, select a minimum of two lines present in introns or exons, where most likely a 'knockout' of the gene will occur (*see* **Note 2**). Questions about their website nomenclature can be found at signal.salk.edu/tdna_FAQs.html.

3.2. Ordering and Preparing Primers for T-DNA Characterization of Mutants

1. The most efficient means to identify homozygous lines is through PCR-based screening. Primers for each SALK line can be obtained throughtheir website (signal.salk.edu/tdnaprimers.2.html), which is found on their SIGnAL iSect toolbox webpage (signal.salk.edu/isects.html). Information on the T-DNA primer design website is imperative to read,

since it contains a detailed description of the T-DNA primer design protocol. Primers for up to 96 T-DNA lines can be obtained at one time – it is recommended that results be saved in an excel spreadsheet. Three additional primers should be ordered for the identification of homozygous lines (Lba1, Lbb1, and Lbc1; listed above).

Alternatively, one can design their own primers through the 'Primer3' automated-primer design website (frodo.wi.mit.edu/cgi-bin/primer3/primer3_www.cgi) – recommended primer design website. The best way to go about doing this is, first, to BLAST the sequence flanking the left border of the T-DNA. To obtain this sequence, return to the '*Arabidopsis* Gene Mapping Tool' webpage (signal.salk.edu/cgi-bin/tdnaexpress). Input your SALK line into the appropriate search box and submit. A graphical representation will be displayed, whereby you then click on your SALK line and a new webpage will come up with information associated only with that SALK line. The left-border flanking sequence is obtained through the link '[seq]'. Copy the sequence and paste it into NCBI's nucleotide BLAST website (www.ncbi.nlm.nih.gov/BLAST/Blast.cgi?PAGE=Nucleotides&PROGRAM=blastn&MEGABLAST=on&BLAST_PROGRAMS=megaBlast&PAGE_ TYPE=BlastSearch&SHOW_DEFAULTS=on). Make sure to specify the Database as 'nr' to include the entire sequence database. In the results section, look for a Blast hit that is associated with a BAC clone and note the starting nucleotide position on the BAC clone. Click on the link for the BAC clone, scroll to the end of the webpage where the entire BAC sequence is listed, and locate the starting nucleotide position that you identified on the previous webpage. Copy approximately 750 bases up and downstream of this position and paste it into the 'Primer3' webpage. On this webpage, there are a number of parameters that you can specify, one of them being the 'Product Size Range', and change it to '900–1100'. This program will pick the optimal set of primers within that range of which you can then order.

2. Upon receipt of primers, make 'stock' and 'working' primer solutions. A recommended primer stock concentration is 100 μM. The simplest way of doing this is to multiply the nmoles of primer by 10 and re-suspend lyophilized primer in this volume (μl) of water. Prepare a working solution of primer at 10 μM.

3.3. Genomic DNA Isolation

The initial goal is to isolate genomic DNA. Grow up a minimum of 8–16 plants on an MS plate – more may be needed but this is sufficient to start. Let the plants reach a stage of growth such that you can isolate an index fingernail worth of tissue (i.e., rosette leaf) into a 1.5 ml eppendorf tube (label the tubes and corresponding plants appropriately), then freeze in liquid nitrogen; store at –20°C if needed. One of the preferred methods for isolating genomic DNA from plant tissue is to use CTAB. Genomic DNA isolated using the CTAB protocol is very clean and reliable and will give very consistent results over multiple freeze-thaw cycles. Alternative methods for isolating genomic DNA can be used; however, it is advised that one can start with this protocol, and can then judge others accordingly. One such alternative will be given that yields consistent and reliable results while being quicker and easier to do.

3.3.1. 'CTAB Buffer' Protocol

1. This step should be done fresh before each use of the protocol. Add 1 μl of β-mercaptoethanol for every 250 μl of CTAB buffer – each sample receives 250 μl of CTAB buffer after grinding. In addition, keep the samples in liquid N_2 and take each one out just prior to use. If allowed to thaw before grinding, it will make it more difficult to solubilize the tissue. Also, set and start your water bath or incubator at 60–65°C and prepare a 70% ethanol solution.
2. Remove eppendorf tube from liquid N_2, grind tissue first for a few seconds, then add 250 μl of CTAB buffer and continue to grind until there are only 'small pieces' of tissue inside of a green solution. After completing up to 24 samples at room temperature, place at 60–65°C for 30–60 min.
3. After the incubation period, add 250 μl of chloroform in the fumehood to each sample, and gently vortex for 3–5 min (it is recommended to use a vortexing platform that can hold a minimum of 24 samples), followed by centrifugation at 10,000 *g* for 4 min.
4. Carefully remove eppendorf tubes from the centrifuge. There will be two layers of solution. Remove 150 μl of the upper aqueous layer, put it into a fresh eppendorf tube and then add 150 μl of isopropanol. Mix well and let it sit for a minimum of 1 h at room temperature or at –20°C. It can also be stored overnight at –20°C without affecting the extraction.
5. After incubating in isopropanol, centrifuge at 10,000 *g* for 20 min at room temperature. Pour off the solution and depending on the size of the starting material one should see a white pellet. Continue even if a white pellet is not seen –then more likely, there is a small amount of genomic DNA present at the bottom of the tube. To wash the pellet, add 0.5 ml of 70% ethanol, mix, and then carefully remove this liquid without removing the white pellet. Allow the pellet to completely dry at this point. One option is to let the samples sit on your bench overnight while being covered.
6. After the pellet has been dried, re-suspend it in 50 μl of sterile autoclaved water by gently vortexing and store at –20°C until

needed. Don't expect all of the pellet to go into solution. Use 1 μl of the 50 μl for a 20–25 μl PCR reaction.

3.3.2. 'Extraction Buffer' Protocol

1. Similar to the first protocol, keep the samples in liquid N_2 and take each one out just prior to use. Also, set and start your water bath or incubator at 60–65°C and prepare a 70% ethanol solution.
2. Remove eppendorf tube from liquid N_2, grind tissue first for a few seconds, then add 400 μl of 'Extraction' buffer and continue to grind until there are only 'small pieces' of tissue inside a green solution. After completing up to 24 samples at room temperature, place them at 60–65°C for 30–60 min.
3. After the incubation, centrifuge at 10,000 *g* for 5 min at room temperature. Transfer 300 μl to a fresh eppendorf tube. It is probable that you will transfer small bits of tissue; however, it will be without any negative consequence. Add 300 μl of isopropanol, mix well, and incubate for a minimum of 1 h at room temperature or at –20°C.
4. After incubating in isopropanol, centrifuge at 10,000 *g* for 20 min at room temperature. Pour off the solution and one should see a pellet. To wash the pellet, add 0.5 ml of 70% ethanol, mix, and than carefully remove this liquid without removing the pellet. Allow the pellet to completely dry at this point.
5. After the pellet has been dried, re-suspend it in 50 μl of sterile autoclaved water by gently vortexing and store at –20°C until needed. Most of the pellet will not go into solution. Use 1 μl of the 50 μl for a 20–25 μl PCR reaction.

3.4. Screening for Homozygous SALK T-DNA Lines by PCR

1. For each T-DNA line that you have isolated genomic DNA, set up two PCR reactions (information primarily based on the use of primers designed by TAIR). It is recommended that this be set up in a 96 well format; therefore 48 individuals can be screened at once. In the first reaction, use the pair of gene-specific primers that span the insertion site. This reaction should give a product size of approximately 1000 bp. In the second reaction, use a T-DNA left border primer (i.e., Lba1) and the right gene-specific primer. The expected product size from this reaction will most likely range from 400 to 800 bp. Alternatively, one can add all three primers into one tube for each reaction (*see* **Note 3**).
2. First, add primers to the PCR reaction tubes for the corresponding SALK T-DNA lines being tested; alternating tubes would have left gene-specific/right gene-specific and Lba1/right gene-specific for each given line. After all primers have been added, make a 'PCR master mix' corresponding to the

total number of reactions and add this to your reactions. Following this, add 1 μl of your genomic DNA preparation. Given that the maximum product size should be approximately 1 kb, an extension time of 1–1.5 min should be sufficient for this reaction. Run your reactions on a 1% agarose gel for 30 min at 100 V. If you are using your primers for the first time, make sure to include a wild-type control so that one can test the efficacy of your primers. Typically in a case where you observe a heterozygous plant for the T-DNA insertion (i.e., a wild-type band from the gene specific primers and a band from the T-DNA left border/right gene-specific primers), both bands should be of equal intensity (*see* **Note 4**).

3.5. Further Characterization of Homozygous T-DNA Line

Probably one of the biggest concerns with T-DNA insertions lines is that there is an average of 1.5 insertions per T-DNA mutant, and that there are often complex integration patterns of T-DNA. Ideally, one should attempt to isolate two or more independent mutants. The probability that both mutants have a second T-DNA and/or other mutation that disrupts the same non-target gene is negligible. If this is not an option after having discovered a worthy phenotype, backcross immediately and test the co-segregation of the T-DNA with your phenotype. Out-crossing to another ecotype, such as Lansberg erecta, especially if it is a desirable phenotype, is also recommended in case mapping is needed to identify the mutation. Additional experiments to provide more definitive proof that the insertional mutation cause the phenotype include, testing for the mRNA transcript by RT-PCR, sequencing the right and left border T-DNA junctions, and complementing the mutant by transformation with a wild-type cDNA or gene. A genomic clone is preferable since it is more likely to replicate the correct pattern and level of gene expression.

3.6. Growing Up the Collection

Once a homozygous line has been identified it needs to be propagated for future experiments. If the plants were screened off of MS plates, transfer the homozygous or heterozygous plants from each line to 10 cm pots. Alternatively one can transfer individual seedlings to a tray with a 48-pot or 96-pot format (*see* **Note 5**).

4. Notes

1. As mentioned in the introduction, a reverse genetics approach to study a given gene family (e.g., >50 genes) has become much more practical in recent years with the advent

of large mutant 'knockout' populations. The challenge is in the consideration of how best to approach such a project. In the case of SALK T-DNA collection, pre-screening on the selectable marker kanamycin for resistant mutants will narrow down the number of individuals that need to be eventually tested by PCR. Since approximately 75% show kanamycin resistance, it is best not to use all the seed sent from the SALK center with this approach. Alternatively, one can initially test for a phenotype of interest without regard to its genotype. Only after a phenotype is found does one then test for its genotype. This is probably the easiest approach, however, if nothing is found and nothing is gained as a result. This is especially problematic if the gene family requires double mutant analysis, for example. The most laborious and expensive approach is to start PCR-based screening for homozygous mutants immediately, thereafter followed by phenotyping. The primary advantage to this approach is that one has a collection of mutants for additional screens and studies.

2. If no 'best' lines are available, you can look for insertions in the promoter, however, since Arabidopsis has most of its regulatory information within the first 500 based pairs of the start site it is best to look only at insertions in this region.
3. Although the temptation may be to save on resources, caution is given when using a 3-primer PCR setup since, in some cases, it can produce additional bands that can confuse the interpretation of the results.
4. If the wild-type band is much lighter, confirm this with your positive control reaction; otherwise there could be a contamination. In the event that one is only able to obtain a wild-type plant for the T-DNA insertion after testing several lines (e.g., 12), try using another left border primer in your PCR reactions.
5. The only caveat to transferring individual seedlings to a 48-pot or 96-pot tray is that, due to the reduced amount of soil in each pot and close proximity of the pots in the tray, the plants tend to be smaller and overcrowded. While this is fine for seed collection purposes, it is not amenable for phenotyping your plants under optimal growth conditions.

References

1. Ferreira, F. J., Kieber, J. J. (2005) Cytokinin signaling. *Curr Opin Plant Biol* 8, 518–525.
2. Bergmann, D. C., Lukowitz, W., Somerville, C. R. (2004) Stomatal development and pattern controlled by a MAPKK kinase. *Science* 304, 1494–1497.
3. Ausubel, F. M. (2002) Summaries of national science foundation-sponsored arabidopsis 2010 projects and national science foundation-sponsored plant genome projects that are generating *Arabidopsis* resources for the community. *Plant Physiol* 129, 394–437.

4. Alonso, J. M., Stepanova, A. N., Leisse, T. J., Kim, C. J., Chen, H., Shinn, P., et al. (2003) Genome-wide insertional mutagenesis of *Arabidopsis thaliana*. *Science* 301, 653–657.

5. Page, D. R., Grossniklaus, U. (2002) The art and design of genetic screens: *Arabidopsis thaliana*. *Nat Rev Genet* 3, 124–136.

Chapter 2

Visualizing Auxin Transport Routes in *Arabidopsis* Leaf Primordia

Danielle Marcos and Thomas Berleth

Abstract

The phytohormone auxin plays a pivotal role in plant development, regulating a myriad of processes including embryo patterning, root patterning, organ initiation, and vein patterning. Auxin is unique among the plant hormones as it is actively transported from cell to cell in a polar fashion. It has recently been discovered that polar auxin transport generates dynamic, local auxin gradients within plant tissues that appear to provide positional information in patterning processes. Visualization of apparent auxin transport patterns has largely been facilitated by the recent creation of translational fusions of GFP to members of the *Arabidopsis* (At)PIN family of auxin efflux associated proteins. Confocal visualization of these fusion products (PIN:GFPs) enables the tracking of apparent auxin transport patterns in a huge number of samples. This visualization method can be combined with experimental interference, such as local auxin application and inhibition of auxin transport, to deduce possible self-organizing auxin-dependent patterning mechanisms and to make them amenable to mathematical modeling.

Key words *Arabidopsis*, *AtPIN* genes, polar auxin transport, leaf development, vascular patterning, confocal microscopy, auxin application, indole-3-acetic acid (IAA).

1. Introduction

The plant hormone auxin is involved in a wide array of plant processes, including embryo patterning *(1)*, photo- and gravitropism *(2)*, phyllotaxis *(3)*, vein patterning *(4)*, and root patterning *(5)*. Auxin distribution within the plant depends on auxin efflux from plant cells, which is tightly linked to the expression and subcellular localization of members of the PIN family of transmembrane proteins *(6)*. In plant cells, PIN proteins cycle rapidly and continuously between an endosomal compartment and the basal cell membrane *(7, 8)*, making possible the rapid relocalization of PIN proteins in response to internal and external cues *(9)*.

Sean Cutler, Dario Bonetta (eds.), *Plant Hormones: Methods and Protocols, Second Edition, vol. 495*

DOI 10.1007/978-1-59745-477-3_2 Springerprotocols.com

Historically, elucidating auxin's precise role in these developmental processes has been difficult due to the difficulty of visualizing auxin directly. However, due to the recent genesis of PIN:GFP translational fusion proteins, one can infer the path and direction of auxin transport from the expression pattern and subcellular localization of PIN:GFPs. Auxin transport routes determined through PIN1:GFP expression patterns are consistent with those determined by other methods, but the use of PIN:GFPs allows one to trace these routes in temporal sequence in huge numbers of samples *(3, 10)*. Recently, this approach has been very successfully combined with live imaging of the shoot meristem to study patterns of auxin transport and gene expression during shoot organ initiation *(11)*. Research involving PIN translational fusions suggests that PIN-dependent auxin distributions are involved in the positioning of lateral organs of the plant root and shoot *(10)*, and the specification of veins of all orders in the plant leaf *(4)*. PIN translational fusions have also been instrumental in revealing the key role of PIN relocation and degradation in root gravitropism *(11)*. All these studies relied on confocal microscopy to examine PIN:GFP expression and subcellular localization in their organ of interest. This basic approach can be used to analyze the role of auxin transport in any developmental process in *Arabidopsis*. Because of high background fluorescence as well as the shape and thickness of the sample, visualization of PIN1:GFP expression in leaf primordia during normal development is particularly challenging. Here, we describe the visualization of auxin-transport patterns during normal leaf development, as well as methods to manipulate those by local auxin application and auxin transport inhibition.

2. Materials

2.1. Seedling Culture

1. Sterilization solution: 15% (v/v) commercial bleach, 0.01% (v/v) Triton X-100 (VWR, West Chester, PA) in sterilized, double-distilled water.
2. Growth medium: Half-strength Murashige and Skoog salts (Sigma, St Louis, MO, USA), 0.5 mg/l morpholino ethane sulfonic acid (MES, Sigma), 0.8% (w/v) agar (BioShop Canada Inc., Burlington, ON), pH 5.7. Note that sucrose is not added to the Growth medium.

2.2. Local Auxin Application

1. IAA paste: up to 10% (w/v) indole-3-acetic acid (IAA; Sigma) mixed in pre-warmed paste consisting of lanolin (Sigma) and a liposoluble red dye (Procter & Gamble; *see* **Note 2**).

2. Insect pins (0.1 mm in diameter; Fine Science Tools, Inc., North Vancouver, BC, CA).
3. Ophthalmic surgical pin holder (Fine Science Tools, Inc.).

2.3. Auxin Transport Inhibition

1. NPA stocks: 1, 10, and 100 mM stock solutions of 1-*N*-naphthylphthalamic acid (NPA; Chem Service).
2. Growth medium (*see* **Section 2.1**).

2.4. Confocal Visualization

1. Double-distilled, sterilized water for mounting (*see* **Note 1**).
2. Microscope slides (76 × 26 mm, VWR).
3. Coverslips (22 × 50 mm, 0.16–0.19 mm thick, VWR).
4. PrecisionGlide needles (30½ gauge, Becton Dickson & Co., Franklin Lakes, NJ, USA).
5. 3 mL Syringe (Becton Dickson & Co.).

3. Methods

The following experiments rely on confocal microscopy [described in detail in ref. *(4)*] to follow PIN:GFP expression and subcellular localization during normal and experimentally challenged leaf primordium development. PIN1:GFP, to which this detailed description specifically refers, is a functional PIN1 protein that is relatively weakly expressed in *Arabidopsis* leaf primordia *(10)*. Experimental manipulations can make PIN1:GFP even more difficult to detect. To obtain high quality images, it is necessary to optimize the following: (a) sample quality – the healthiest seedlings tend to produce the best quality images; (b) mounting – the leaf primordium should be mounted as flat as possible, in the correct orientation, to minimize the number of tissue layers that the light must penetrate; (c) visualization parameters – while others' suggestions help, these must ultimately be determined empirically for each marker and confocal microscope; (d) viability of the sample – samples were routinely visualized immediately after mounting as PIN1:GFP shows signs of decreased expression and relocalization in response to treatment within 5–10 min. Once the sample has been mounted, it is necessary to work quickly and without interruptions. Therefore, optimized parameters and a routine protocol must be established first that can then be used to collect data from large numbers of samples.

3.1. Seedling Culture

1. Leaf synchronization is critical for pattern reproducibility. The following seedling culture protocol incorporates strategies to ensure synchronous leaf initiation and preparation. To

prepare seedlings for confocal visualization, incubate PIN1:GFP seeds *(10)* in 70% ethanol for 1 min and then vortex in sterilization solution for 20 min.

2. Wash 10 times in sterile, double distilled water, incubating for at least 10 min between washes. This incubation period, a modification of the synchronization procedure by Petrov and Vizir *(12)*, allows the seeds to imbibe the water and synchronizes their germination.
3. Sow seeds on growth medium in sterile Petri plates at a density of 1 seed/cm^2. Seal plates and wrap in aluminum foil to exclude light.
4. Stratify seeds at 4°C for 5 days. A 5-day stratification period will provide the best synchronization of germination. Shorter stratification periods will result in progressively less synchronized seedling development.
5. Following stratification, incubate plates in growth chambers with continuous fluorescent light (100 $\mu E/m^2/s^2$) at 25°C.

3.2. Local Auxin Application

1. Preparation of IAA paste: Place 1 mL of lanolin in a microcentrifuge tube and place it in a 65°C heating block, to melt lanolin.
2. Add a small quantity (approximately, 25 μl) of liposoluble red dye (*see* **Note 2**), such that mixture turns a deep red color. If necessary, re-adjust the volume to 1 mL by removing a small amount of the lanolin mixture with a pipette. Reserve lanolin-dye paste to treat controls.
3. Remove the microcentrifuge tube from the heating block and allow it to cool slightly, such that it is still liquid but no longer feels warm to the touch (*see* **Note 3**).
4. Add IAA to a final concentration of up to 10% (w/v). Incubate the microcentrifuge tube at 4°C for 15–30 min, to cool and solidify the paste. Due to the lability of IAA, the IAA paste should be used immediately.
5. Application of IAA paste: Using a small spatula, smear a small amount of IAA paste on a piece of weighing paper. Using dissecting microscope to examine the plate of cultured seedlings, select a healthy seedling of the correct stage to treat (*see* **Note 4**).
6. Holding the insect pin in the pin holder, drag the insect pin through the IAA paste. This will cause the paste to form a long thin peak on the tip of the pin.
7. Working under the dissecting microscope, very carefully touch the tip of this peak to one side of the leaf primordium. A small drop of IAA paste will be transferred onto that side of the primordium (*see* **Note 5**).

8. Treat control seedlings in the same way with lanolin-dye paste that lacks IAA.
9. Return the plate of treated seedlings to the growth chamber.
10. After 24 or 48 h post-treatment, proceed with confocal visualization.

3.3. Auxin Transport Inhibition

1. Preparation of NPA growth medium: Prepare plates of growth medium containing 0, 5, 10 or 50 μM NPA by adding an appropriate volume of 1, 10 or 100 mM NPA stock solution to 25 mL molten, solid growth medium (per plate).
2. Sow and stratify PIN1:GFP sterilized seeds as described in ***3.1, Item 4.***
3. Visualize NPA-treated seedlings at the confocal microscope daily from 2 to 5 DAG.

3.4. Confocal Visualization

1. Mounting of samples: Place one seedling on a drop of double distilled, sterile water on a microscope slide.
2. Using the sharp edge of the needle tip as a blade, cut the root above the root-hypocotyl junction and remove. This decreases the thickness of the preparation considerably, enabling better visualization of the leaf primordia. However, without a root, the treatment-induced changes in PIN1:GFP expression will follow 5–10 min post-mounting. Therefore, individual seedlings are imaged immediately after mounting.
3. After removing the root, place one needle on each cotyledon and gently pull them apart, exposing the first leaf primordia that lie between the cotyledon petioles. It may be necessary to slightly damage the cotyledons to force them to lie flat against the slide, as shown in **Fig. 2.1**. If imaging the third leaf or subsequently formed rosette leaves, use the needle to cut away the cotyledons and any leaf primordia covering the

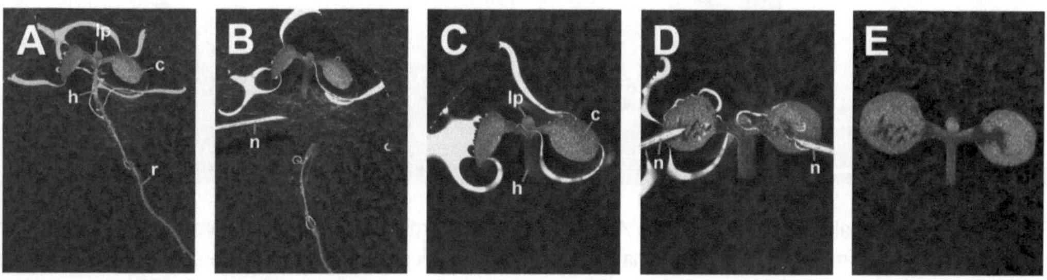

Fig. 2.1. Mounting of *Arabidopsis* seedlings for confocal visualization of the leaf primordia. (c) cotyledon; (h) hypocotyl; (lp) leaf primordium; (n) needle; (r) root. (**A**) Seedling on a drop of water on microscope slide. (**B**) Seedling with hypocotyl cut above the root-hypocotyl junction. (**C**) Seedling prior to flattening. (**D**) Seedling flattened by pulling on the cotyledons with needles. (**E**) Seedling compressed by coverslip.

primordium of interest, and orient the seedling such that this primordium faces up.

4. Place one edge of the coverslip against the slide and, balancing the other edge of the coverslip on the needle, gently and slowly lower the coverslip onto the slide, avoiding the formation of air bubbles (*see* **Note 6**). Sudden dropping of the coverslip onto the preparation will result in damage to the sample.
5. Optimization of visualization settings (*see* **Note** 7): When working with weak fluorescent markers, like the PIN:GFPs, one is typically operating close to the lower detection limit of the microscope. When determining the optimal visualization conditions, begin with all parameters (laser output, laser transmission, pinhole, and gain) at low settings (*see* **Note 8**).
6. Increase gain until an acceptable image is obtained. Only if fluorophore detection is still insufficient, gradually increase the pinhole, as large pinholes will result in a blurred image. If the detection is still unsatisfactory, resort to increasing the transmission and/or output of the excitation laser, keeping in mind that this increases photodamage of the sample.
7. Due to the difference in PIN1:GFP signal intensity in different parts of the leaf primordium, visualization setting that enable the detection of PIN1:GFP in incipient higher order veins will result in total signal saturation in the lateral epidermis, obscuring PIN1 polarity in this area. Take multiple images of the same focal plane, some with higher pinhole/gain settings to capture the weakest areas of PIN1:GFP expression and some with lower pinhole/gain settings to resolve PIN1 localization in areas of very high PIN1:GFP expression (*see* **Fig. 2.2**).

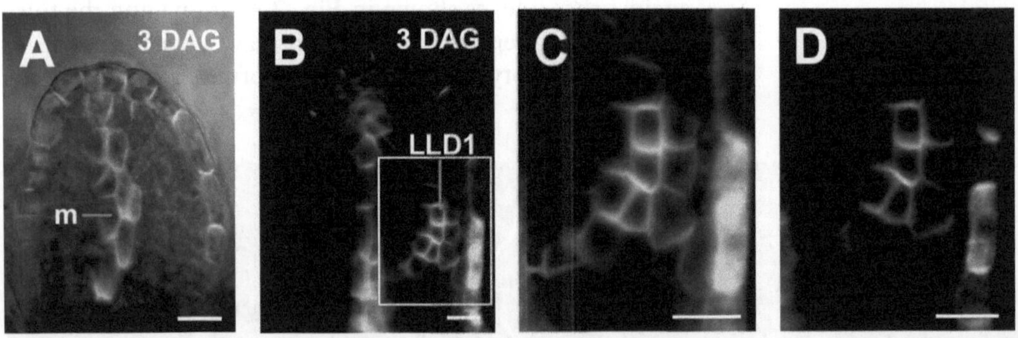

Fig. 2.2. Confocal visualization of PIN1:GFP in *Arabidopsis* leaf primordia. (*Upper right*) Primordium age in days after germination (DAG). (m) midvein; lower loop domain of first (most distal) loop [see ref. *(4)*]. (**A**) Merged DIC and GFP image of a first leaf primordium at 3 DAG, showing PIN1:GFP expression in the midvein and epidermis. (**B**) GFP image of a first leaf primordium at 3 DAG, showing PIN1:GFP expression in the midvein, lateral epidermis, and the lower domain of the first loop. (**C**) Enlargement of the area boxed in (**B**), showing the complete PIN1:GFP expression pattern in the lower loop domain of the first loop. (**D**) Same area shown in (**C**), imaged with a smaller pinhole to visualize PIN1:GFP localization in the areas of highest expression. *Bars*: **(A–D)** 10 μm (Reproduced from ref. *(4)* with permission from CSHL Press).

Those individual images can be used to analyze relative signal intensities. Composite images constructed from single images taken with different settings cannot be used to quantify differences in PIN1:GFP expression, but rather document all polarity features of the PIN1:GFP expression pattern in the developing primordium.

8. Individual images can be used to analyze relative signal intensities using look-up tables (LUTs). The human eye cannot distinguish 256 different levels of brightness, the minimum recorded by all confocal microscopes *(13, 14)*. Therefore, to increase distinction of monochromatic shades, and thus convey the true spectral detection range of the data, critical fluorescence images can be turned into 8-bit gray-scaled images and their intensity expressed as pseudo-colors, using a rainbow spectrum look-up-table (LUT) of ImageJ (National Institutes of Health, http://rsb.info.nih.gov/ij).
9. Since the leaf primordium is not perfectly flat, its complete PIN1:GFP expression pattern cannot be captured in an image of a single focal plane. Cover all features of the expression pattern by taking images at several focal planes, either by focusing manually through the sample and taking images at intervals or by using the z-stack function of the confocal microscope. The resulting images can be compiled and overlaid using microscope software or Adobe Photoshop 7.0 (Adobe Systems). Sample composite images, reconstructed from multiple single images of a IAA-treated and untreated first leaf primordia, are shown in **Figs. 2 & 3**.

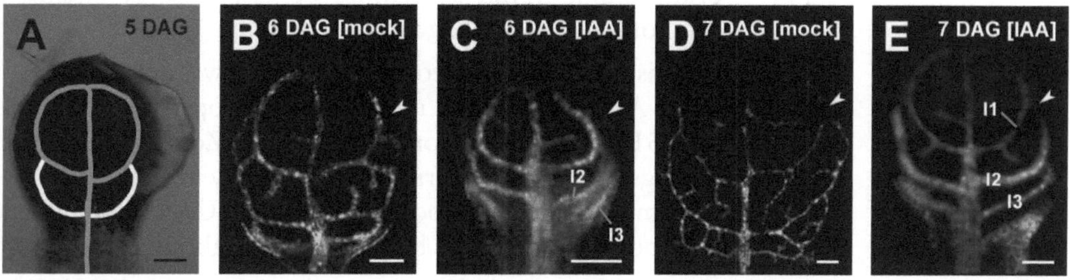

Fig. 2.3. Effects of local auxin application on PIN1:GFP expression in *Arabidopsis* leaf primordia. (*Upper right*) Primordium age in days after germination (DAG) and treatment (IAA/mock, in *square brackets*). (*Yellow arrowheads*) Site of application; (I1, I2, I3) loop 1, loop 2, loop 3 [ordered distal to proximal, see ref. *(4)*]. (**A**) DIC image of a first leaf primordium 5 DAG, showing predicted positions of procambium (*grey*) and marker-expression defined preprocambium (*white*). (**B**) GFP image of a mock-treated first leaf primordium 6 DAG, showing the PIN1:GFP expression in the midvein, in three loops and in higher order veins. (**C**) GFP image of a IAA-treated first leaf primordium 6 DAG, showing increased PIN1:GFP expression in the area of the third loop on the treated side. (**D**) GFP image of a IAA-treated first leaf primordium 7 DAG, showing an extra fourth loop in the PIN1:GFP pattern on the treated side. (**E**) *Bars*: (**A–E**) 50 μm [Reproduced from ref. *(4)* with permission from CSHL Press].

4. Notes

1. Because viscosity is not an issue, distilled water rather than glycerol solutions can be used as a mounting medium.
2. Any brightly colored lipstick could be used as a liposoluble red dye.
3. IAA is heat labile. Therefore, to prevent degradation, the lanolin mixture must be cooled until it no longer feels warm to the touch. Once cooled, IAA is added immediately, because the lanolin paste solidifies quickly at this temperature.
4. To study the effect of local auxin application on loop formation, one would ideally treat those first 4 DAG leaf primordia, in which the second and third loops have not yet formed as preprocambium *(15)*. However, it is very difficult to apply IAA paste unilaterally on such tiny primordia. A preferred stage would therefore be 5 DAG, a stage when the third loop has not yet formed as preprocambium *(15)*. At 4 DAG, the two first leaf primordia lie close together, pointing straight up when viewed from above. By 5 DAG, in contrast, the first leaf primordia have separated and lie perpendicular to the axis of the incoming light.
5. The size of the drop will depend on the width of the IAA paste peak on the pin. Practice drawing the pin through the lanolin paste, to create the thinnest peak possible. This peak can then be used to treat multiple seedlings.
6. If an air bubble forms in the slide preparation, but it does not obscure the leaf primordium, the slide can be used without difficulty. If an air bubble forms over the area of the leaf primordium, it can sometimes be displaced by adding a drop of water to the edge of the coverslip.
7. The visualization protocol described here was developed on a Zeiss Axiovert 100 M microscope equipped with a Zeiss LSM510 laser module confocal unit (Carl Zeiss). The optimal settings have to be determined empirically for other fluorescent markers and confocal microscopes. Once determined, the optimal settings can be re-used throughout a study.
8. On the Zeiss LSM510 confocal microscope, PIN1:GFP can be visualized using GFP was the 488-nm line of an Argon laser at 50–55% of laser output and 4–10% transmission, and with a 505–530-nm band-pass filter. The gain used is 3 (100% of maximum gain) and the pinhole is between 5 and 14 Airy units (36–100% maximum pinhole using 40 × Plan Neofluar oil-immersion objective with a numerical aperture of 1.3.)

Acknowledgments

The authors would like to thank Jiří Friml for PIN1:GFP seeds, and Enrico Scarpella for his helpful comments on this manuscript. This work was supported by Discovery Grants of the Natural Sciences and Engineering Research Council of Canada (NSERC) (to T.B.) and by an Ontario Premier's Research Excellence Award (to T.B). D.M. was supported by an NSERC PGS-M fellowship.

References

1. Weijers, D. and Jurgens, G. (2005) Auxin and embryo axis formation: the ends in sight? *Curr Opin Plant Biol* 8, 32–7.
2. Swarup, R., Kramer, E. M., Perry, P., Knox, K., Leyser, H. M., Haseloff, J., Beemster, G. T., Bhalerao, R. and Bennett, M. J. (2005) Root gravitropism requires lateral root cap and epidermal cells for transport and response to a mobile auxin signal. *Nat Cell Biol* 7(11), 1057–65.
3. Reinhardt, D., Pesce, E. R., Stieger, P., Mandel, T., Baltensperger, K., Bennett, M., Traas, J., Friml, J. and Kuhlemeier, C. (2003) Regulation of phyllotaxis by polar auxin transport. *Nature* 426, 255–60.
4. Scarpella, E., Marcos, D. and Berleth, T. (2006) Control of leaf vascular patterning by polar auxin transport. *Genes Dev* 20(8), 1015–27.
5. Blilou, I., Xu, J., Wildwater, M., Willemsen, V., Paponov, I., Friml, J., Heidstra, R., Aida, M., Palme, K. and Scheres, B. (2005). The PIN auxin efflux facilitator network controls growth and patterning in *Arabidopsis* roots. *Nature* 433, 39–44.
6. Paponov, I. A., Teale, W. D., Trebar, M., Blilou, I. and Palme, K. (2005) The PIN auxin efflux facilitators: evolutionary and functional perspectives. *Trends Plant Sci* 10, 170–7.
7. Friml, J., Benkova, E., Blilou, I., Wisniewska, J., Hamann, T., Ljung, K., Woody, S., Sandberg, G., Scheres, B., Jurgens, G. and Palme, K. (2002a) AtPIN4 mediates sink-driven auxin gradients and root patterning in *Arabidopsis*. *Cell* 108, 661–73.
8. Geldner, N., Friml, J., Stierhof, Y. D., Jurgens, G. and Palme, K. (2001) Auxin transport inhibitors block PIN1 cycling and vesicle trafficking. *Nature* 413, 425–8.
9. Friml, J., Wisniewska, J., Benkova, E., Mendgen, K. and Palme, K. (2002b) Lateral relocation of auxin efflux regulator PIN3 mediates tropism in *Arabidopsis*. *Nature* 415, 806–9.
10. Benkova, E., Michniewicz, M., Sauer, M., Teichmann, T., Seifertova, D., Jurgens, G. and Friml, J. (2003) Local, efflux-dependent auxin gradients as a common module for plant organ formation. *Cell* 115, 591–602.
11. Heisler, M. G., Ohno, C., Das, P., Sieber, P., Reddy, G. V., Long, J. A., Meyerowitz, E. M. (2005) Patterns of auxin transport and gene expression during primordium development revealed by live imaging of the *Arabidopsis* inflorescence meristem. *Curr Biol* 15(21), 1899–911.
12. Abas, L., Benjamins, R., Malenica, N., Paciorek, T., Wisniewska, J., Moulinier-Anzola, J.C., Sieberer, T., Friml, J. and Luschnig, C. (2006) Intracellular trafficking and proteolysis of the *Arabidopsis* auxin-efflux facilitator PIN2 are involved in root gravitropism. *Nat Cell Biol* 8(3), 249–56.
13. Petrov, A. P. and Vizir, I. Y. (1982) On seed germination synchronization of Arabidopsis thaliana (L.) Heynh. *Arabidopsis Information Service* 19.
14. Russ, J. C. (1995) The image processing handbook. CRC Press Boca Raton, FL.

15. Young, I. T., Gerbrands, J. J. and van Vliet, L. J. (1998) Fundamentals of image processing. In The digital signal processing handbook (eds. V.K. Madisetti and D.B. Williams), pp. 51.01–51.81. CRC Press, Boca Raton, FL.

Chapter 3

Gene Expression Analyses for Elucidating Mechanisms of Hormonal Action in Plants

George W. Bassel and Nicholas J. Provart

Abstract

Analysis of large-scale gene expression data sets is proving to be a powerful tool for gene function prediction, *cis*-element discovery and hypothesis generation using *Arabidopsis thaliana*. Public initiatives led by the AtGenExpress Consortium and experiments conducted by individual researchers to document the transcriptome of *Arabidopsis thaliana* have led to a large numbers of data sets being made publicly available for data mining by so-called "electronic northerns", co-expression analysis and other means. Given that approximately 50% genes in *Arabidopsis* have no function ascribed to them by "traditional" homology searches, and that only around 10% of the genes have had their function confirmed in the laboratory, these analyses can accelerate the identification of potential gene function with a mouse-click. This chapter covers the use of data mining tools available at the Bio-Array Resource (www.bar.utoronto.ca) for hypothesis generation in the context of plant hormone biology.

Key words: Gene expression analysis, co-expression analysis, data mining, *cis*-element prediction, microarray, gene function prediction, *Arabidopsis thaliana*.

1. Introduction

This chapter will touch on four aspects of online gene expression analysis: (1) generating "electronic northerns", (2) the eFP Browser (electronic Fluorescent Pictograph Browser) to graphically examine the expression pattern of one or several genes across many data sets, (3) Expression Angler to identify co-expressed genes and (4) Promomer to predict potential *cis*-elements in individual promoters or in groups of promoters.

The "guilt-by-association" paradigm is based on the principle that genes involved in the same pathway will share a common similar gene expression pattern. This approach for assigning gene

Sean Cutler, Dario Bonetta (eds.), *Plant Hormones: Methods and Protocols, Second Edition, vol. 495*

DOI 10.1007/978-1-59745-477-3_3 Springerprotocols.com

function to uncharacterised genes is a well-recognised method in yeast, and has also been extended to multicellular organisms, such as human and mouse *(1–3)*. In the case of *Arabidopsis*, thousands of data sets have been generated by individual laboratories and by the AtGenExpress Consortium including the Developmental Map data set *(4)* and Global Stress data set *(5)* and others as yet unpublished. These have been archived in public microarray databases such as the Gene Expression Omnibus – GEO *(6)*, ArrayExpress *(7)*, The Arabidopsis Information Resource – TAIR *(8)*, NASCArrays *(9)*, Genevestigator *(10)* and the Bio-Array Resource – the BAR *(11)*. All of these sites have query pages for exploring the data sets in order to obtain the expression levels for a desired gene under specific treatments or in given tissues. In addition, several web-based co-expression analysis tools for identifying genes that are co-expressed with a researcher's favourite gene have been developed in the past couple of years. Examples of such programs are ATTED-II *(12)*, the Arabidopsis Co-expression Tool – ACT *(13)*, AthCoR *(14)* and Expression Angler *(11)*. The specific features of each of these programs are described thoroughly by Aoki and colleagues, who also discussed the topic of co-expression networks in plants *(15)*.

There have also been more computational studies that use large-scale expression data sets to infer gene function and regulatory modules, particularly in the context of curated lists of genes. Successful studies using this approach include those for isoprenoid biosynthesis *(16)* or for cytochrome P450s *(17)*. Such large-scale studies will be ignored here in favour of a focusing on the use of web-based tools for identifying expression patterns and co-expressed sets of genes for one's gene or genes of interest, especially in the context of hormonal action in plants.

As an extension of the above, genes that are co-expressed often contain common *cis*-regulatory elements in their promoters. This chapter also discusses *cis*-element enrichment analyses and *de novo* discovery aspects using publicly-available tools.

2. Materials

For web-based analysis, all that is required for the methods described below is access to a computer with good internet connection and a web browser. Data from the tools that are described below may be downloaded as text-based files that can be opened with a spreadsheet program, such as Microsoft Excel, for further manipulation. Gene lists, in the sense of a series of Arabidopsis Genome Initiative (AGI) identifiers such as At3g24560, can be copied from a spreadsheet, whereby each gene identifier should be in its own cell and on a different row of the spreadsheet. Alternately the list may be copied from a word processing program or simple text editor, again such that each AGI identifier is on its own line.

3. Methods

3.1. Electronic Northern Analysis – eFP Browser

The electronic Fluorescent Pictograph (eFP) Browser is a tool for the analysis and visualisation of microarray data *(18)*. Typically with other tools, a two dimensional matrix of colours representing different levels of transcript abundance – a so-called heatmap – has been used to visualise the expression levels of several genes across many experiments. In contrast, the eFP Browser contains pictographic representations of the organism, tissue, cell type or condition examined for a given set of experiments. Visual representations facilitate user interpretation of these data over the coloured squares used in "traditional" heatmaps.

1. Point a web browser to www.bar.utoronto.ca and select the "Arabidopsis eFP Browser" link on the BAR homepage.
2. Select the "Data Source" to visualise different compendia of microarray data. The default is the Arabidopsis Developmental Map from Detlef Weigel's group in Tübingen, Germany *(4)*.
3. Enter the AGI identifier in the "Primary AGI ID" field for a given gene of interest. The Mode may be left in the default "Absolute" setting. A discussion of the various Modes may be found in the Notes section (*see* **Note 1**).
4. Click GO.
5. Standard deviation filtering may be clicked in the output screen to mask samples with large variability between replicates. These samples will be coloured grey in the eFP Browser output.

The correct interpretation of the eFP Browser output is contingent on the colour scale depicting raw signal values obtained from the microarray data sets. The colour scale in the example output image shown in **Fig. 3.1** is found in the bottom left of the output image for the Developmental Map and ranges from a lighter yellow colour (negligible expression level) to a darker red colour (higher expression level). In the case of the normalisation method used for public microarray data in the BAR, a raw signal value below 20 is background; genes with maximum signal intensity lower than this value are not significantly expressed. *CYP707A2* in **Fig. 3.1** has a maximum value of 336 in the developmental data set, allowing for the meaningful interpretation of these

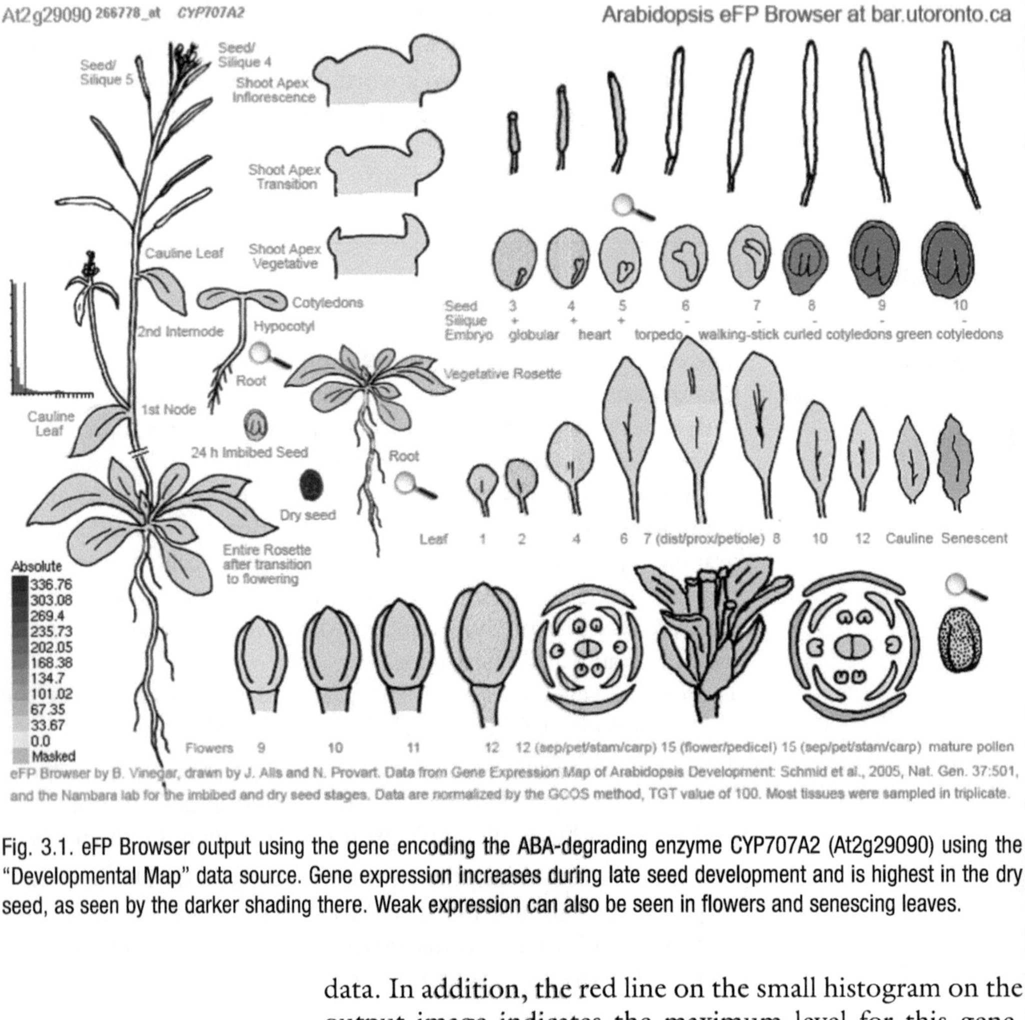

Fig. 3.1. eFP Browser output using the gene encoding the ABA-degrading enzyme CYP707A2 (At2g29090) using the "Developmental Map" data source. Gene expression increases during late seed development and is highest in the dry seed, as seen by the darker shading there. Weak expression can also be seen in flowers and senescing leaves.

data. In addition, the red line on the small histogram on the output image indicates the maximum level for this gene, with the average expression level distribution of all other genes indicated by the grey bar graph – the small ticks denote steps of 100 units while the larger ticks denote steps of 1,000 expression units: clicking on this histogram will enlarge it. In all of the compendia in the eFP Browser, roughly 50% of the genes are expressed at 100 expression units or lower.

By placing the mouse pointer over a given sample, the signal intensity and standard deviation for each sample can be seen in a small pop-up window. Clicking on the sample links to the original data source.

6. Once an AGI ID has been entered, the user can view other data set compendia by changing the option in the "Data Source" pull down menu.
7. **Figure 3.2** represents the same gene as in **Fig. 3.1** using the "Hormone" Data Source. It is readily apparent that this gene is rapidly and specifically induced by ABA treatment.

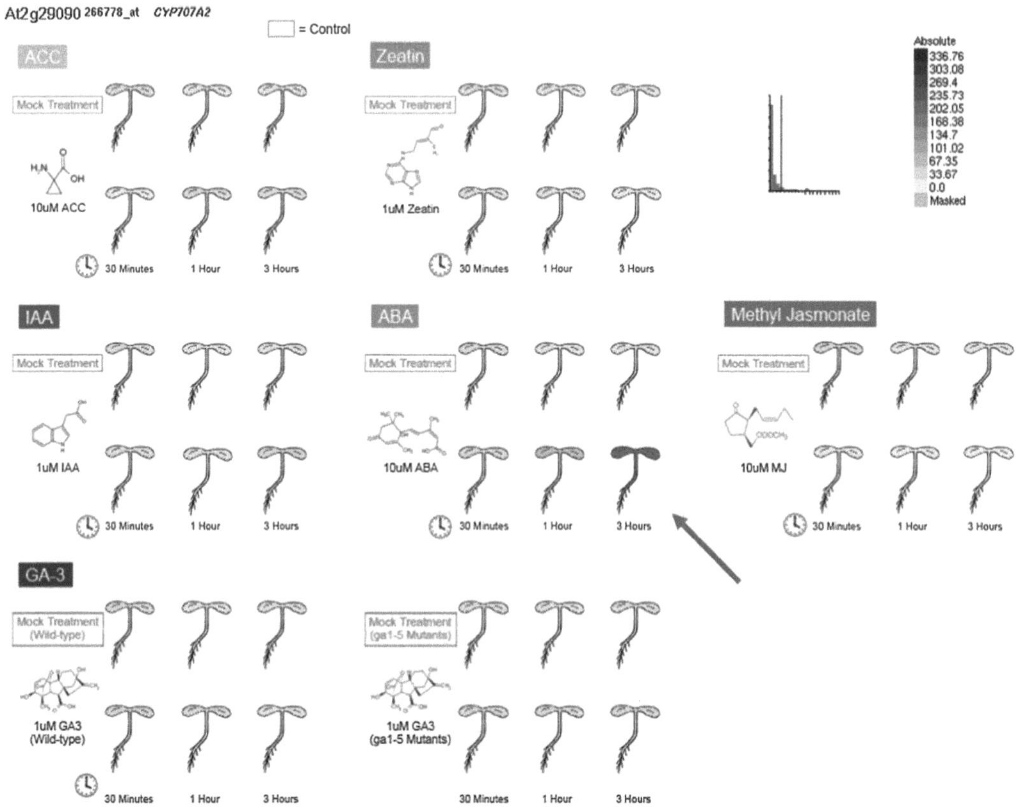

Fig. 3.2. eFP Browser output for *CYP707A2* (At2g29090) using the "Hormone" Data Source. This gene is specifically and rapidly induced by ABA in Arabidopsis seedlings, as indicated by the darker shading, pointed to by the *arrow*.

3.2. Electronic Northern Analysis – Expression Browser

The Expression Brower can be considered an extension of the eFP Browser, accepting lists of genes rather than single AGI identifiers. The Expression Browser's output for gene expression is in the form of a heatmap matrix where each row is indexed by a given AGI identifier and each column specifies the sample from which the data were generated. Across the top of the heatmap, output are multiple fields with different colours representing information about the sample, including tissue type, whether the sample is a treatment or control, if the data are from a mutant genotype, the relative time scale for time-series experiments, and the age of the plant from which the samples were taken.

In addition to providing information on transcript abundance, the Expression Browser also performs hierarchical clustering in order to group AGI IDs with similar patterns of gene expression in the chosen samples, and displays both documented and predicted interactions between the protein products of the genes in question. These data are displayed to the right of the heatmap.

1. Point a web browser to www.bar.utoronto.ca and select the "e-Northerns with Expression Browser" link on the BAR homepage.
2. Select the category of data sets you wish to query. Only one category may be selected at a time.
3. Data sets within the data category may be subselected in the "Research Area" window. Multiple areas may be selected at once by holding down the Ctrl key. Not clicking in this area selects all experiments.
4. Tissue types, growth stages and time points may be subselected in a similar fashion to **Step 3**.
5. The default output option is the "Average of replicate treatments relative to average of appropriate control". This setting is recommended. However, the second option "Average of replicate treatments" is also useful when considering absolute expression levels.
6. The list of AGI IDs may then be pasted into the appropriate text field below. Each AGI ID must be on its own line.
7. Click Submit.
8. On the output page, links to multiple options for displaying the query results are then displayed. The two options in bold represent the results which have undergone clustering, and are recommended. The first of the two represents the raw ratio measurements for the genes. These ratios will range from 0 to 1 for genes that are down-regulated relative to levels from a control set of experiments to greater than one where the genes are upregulated (in the case of the Developmental Map, Tissue Specific, and the Seed series, the control for each gene is its median expression level across all the samples displayed; in the case of abiotic and biotic treatments, the control level for a gene is the level in the mock-treated or untreated plants). In the second output link, the data have been $\log_2$ transformed. Log-transformed data are useful in highlighting differences between data points more clearly than the absolute values.

   Scrolling the mouse over the heatmap displays information about that particular sample in fields above the heatmap. The "Description" field provides a text description of the sample. Clicking on the sample column in the heatmap provides a link to the original data. The maximum colour scale for the heatmap can be adjusted in the textbox beside the "Max" button. This can aid in visualising differences between genes expressed at levels lower than the calculated colour scale maximum. Other click options are discussed in the Notes section (*see* **Note 3**).

   To the right of the heatmap is a tree-like graphic representing the way in which the data have been hierarchically clustered. In **Fig. 3.3**, ACS1 and ACS4 show the most similar expression responses: the Pearson correlation coefficient, which can be read

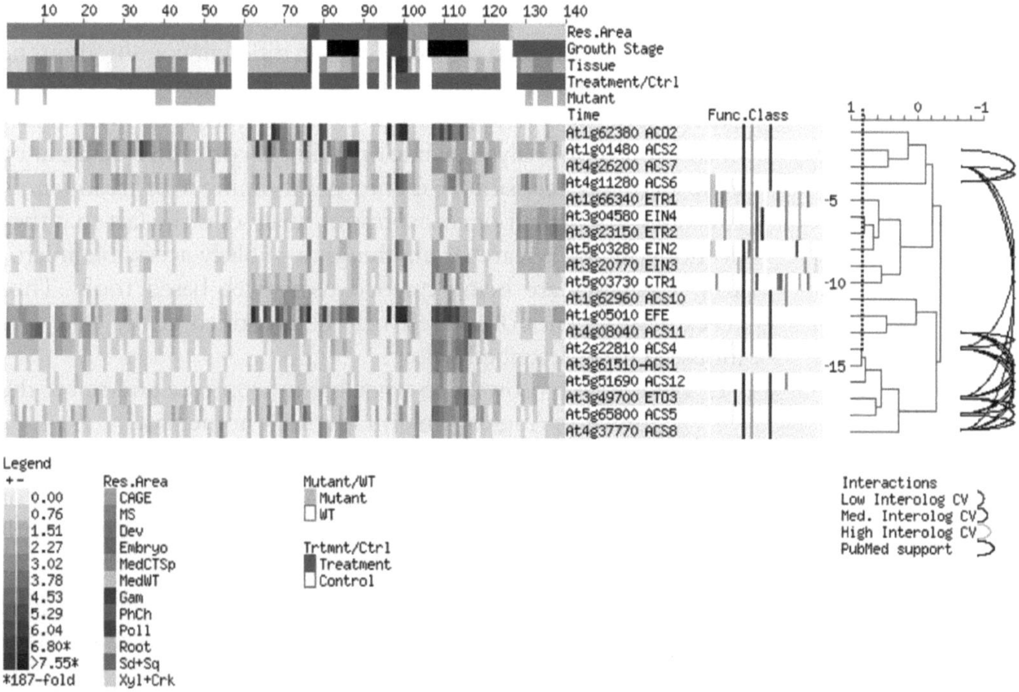

Fig. 3.3. Expression Browser output for a log-transformed and clustered list of genes involved in ethylene biosynthesis and response using the Extended Tissue Series data set. The clustering tree and protein interactions can be seen to the *right* of the heatmap. ACS1 and ACS4 show the most similar expression response patterns, but have not been documented to interact in the literature. The dotted line has been added to show how to use the scale above the tree to read off a correlation score between ACS1 and ACS4 (see text).

from the small scale above the tree, is around 0.8 (a value of 1 indicates identical responses between the two genes, a value of 0 is no similarity in response and a value of −1 indicates that the responses are opposite). Some caution in interpreting the meaning of these results is advised (*see* **Note 2**). The arcing lines represent proven and putative interactions between proteins, as indicated in the legend below the heatmap.

3.3. Co-expression Analysis – Expression Angler

Co-expression analysis can be thought of as a generalisation of the more "traditional" microarray analysis in which the responses of many genes under a treatment or from a specific tissue type are compared to their responses, or lack thereof, in a reference sample, with perhaps a handful of samples or treatments being examined. In such a "traditional" analysis, genes that are upregulated relative to their levels in a reference sample are examined for similar Gene Ontological (GO) categories and unknown genes are ascribed a function based on a similar pattern of response to known genes, as described in the preceding section. A key difference between co-expression analysis and more classical microarray analyses, however, is that co-expression analysis examines a large number of data sets, while classical studies draw information from a relatively smaller number of data sets representing different conditions in the researcher's own experiment. Greater statistical power is generated as a result of larger sample sizes (n=number of data sets in database) in co-expression analyses, while only simple correlative data can be generated in a "classical" experiment (n=number of treatments). An additional benefit of co-expression analyses is the fact that they do not cost the researcher anything, as publicly-accessible microarray databases are being queried using web-based tools. Thus co-expressed genes of interest may be identified at the click of a mouse. This is described below.

1. Point a web browser to www.bar.utoronto.ca and select the "Expression Angler" link on the BAR homepage.
2. The Expression Angler program offers a standard input page and a special input page for selecting specific samples and/or designing a custom bait. These latter aspects will be dealt with in the Notes section (*see* **Notes 4** and **5**, respectively).
3. Enter an AGI identifier in the first box. This identifier will be used to extract the gene expression values associated with that gene in a specified compendium.
4. Either enter an *r*-value cutoff and select the "*r*-value cutoff range" option, or select the "Top 25, 50 or 100 hits" or the "Bottom 25, 50 or 100 hits". The Pearson correlation coefficient (*r*) is used to assess how similar gene expression patterns are between one's gene of interest and all the other gene expression patterns in the compendium in question. Selecting the default cutoff will return genes that are co-expressed, insofar as they meet the cutoff criterion of being the top 25 genes with the best *r*-values to the gene of interest, while one of the other "Top" or "Bottom" options will return genes the specified number of genes, regardless of *r*-value. Those returned by the "Bottom" option are genes that are potentially responding in the opposite manner to one's gene of interest.
5. Select a compendium in which to search. There are currently 8 data sets (or compendia) in which one can search: 1 large set from NASCArray containing almost 400 samples, one set from the Bio-Array Resource in-house expression profiling service and 6 data sets that are collections of data from various AtGenExpress expression profiling experiments. Alternately, one can upload one's own data set.
6. Click Submit and wait about 15 s for the BAR server to generate its output.
7. On the output page, there are a number of links. The first link to "View data set as text" leads to a text-based, tab character-delimited file that one can download to open in a spreadsheet program.

8. The next two links "View formatted data set" and "View formatted data set after median centring and normalisation" activate a visualisation program called Data MetaFormatter that allow one to examine graphical representations of the expression patterns of the genes passing the filtering criteria. This output is similar to that of the heatmap generated by the Expression Browser tool (**Section 3.2**). Both these outputs offer the same gene list information. In the first case, the expression values are not adjusted, while in the second case they are median-centred and normalised on a per gene basis, allowing better appreciation of the features (samples) that contribute most to the co-expression score. This can be seen in **Fig. 3.4**.
9. The next hyperlink in the output is to "Activate the Promomer Program using the AGIs identified". Promomer is a *cis*-element prediction program which is discussed in **Section 3.4**.
10. Several other links to other, external *cis*-element mapping and prediction programs are also offered at the bottom of the Expression Angler output page (*see* **Note 6**).
11. Genes are listed in the output in descending order according to their *r*-value score. Genes may have annotations associated with them, or be annotated as "expressed protein". Using the "guilt-by-association" paradigm, these may also be involved in one's biological process of interest.

3.4. cis-Element Analysis and Prediction

1. Point a web browser to www.bar.utoronto.ca and select the "Promomer" link on the BAR homepage.
2. Promomer offers four different types of promoter analyses. The focus of this section will be on "Promomer 2", which takes as an input a list of AGI IDs and then performs a word

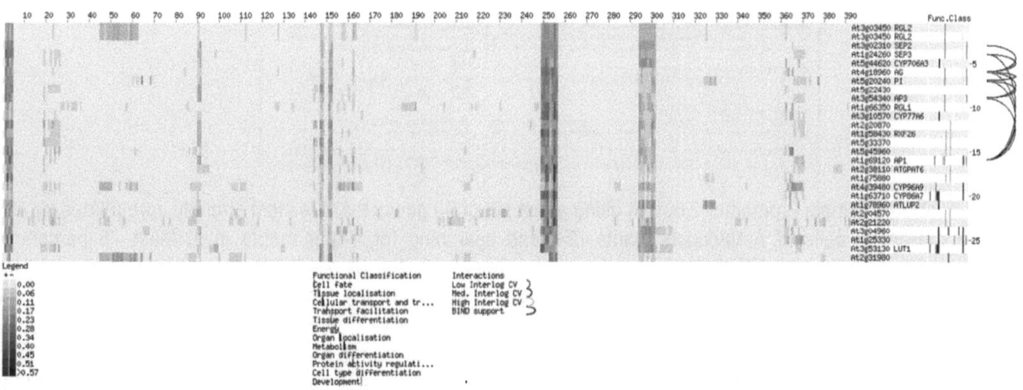

Fig. 3.4. An example heatmap output for a co-expression analysis using *RGL2* and the NASCArrays database. Many floral homeotic genes are returned.

A

B

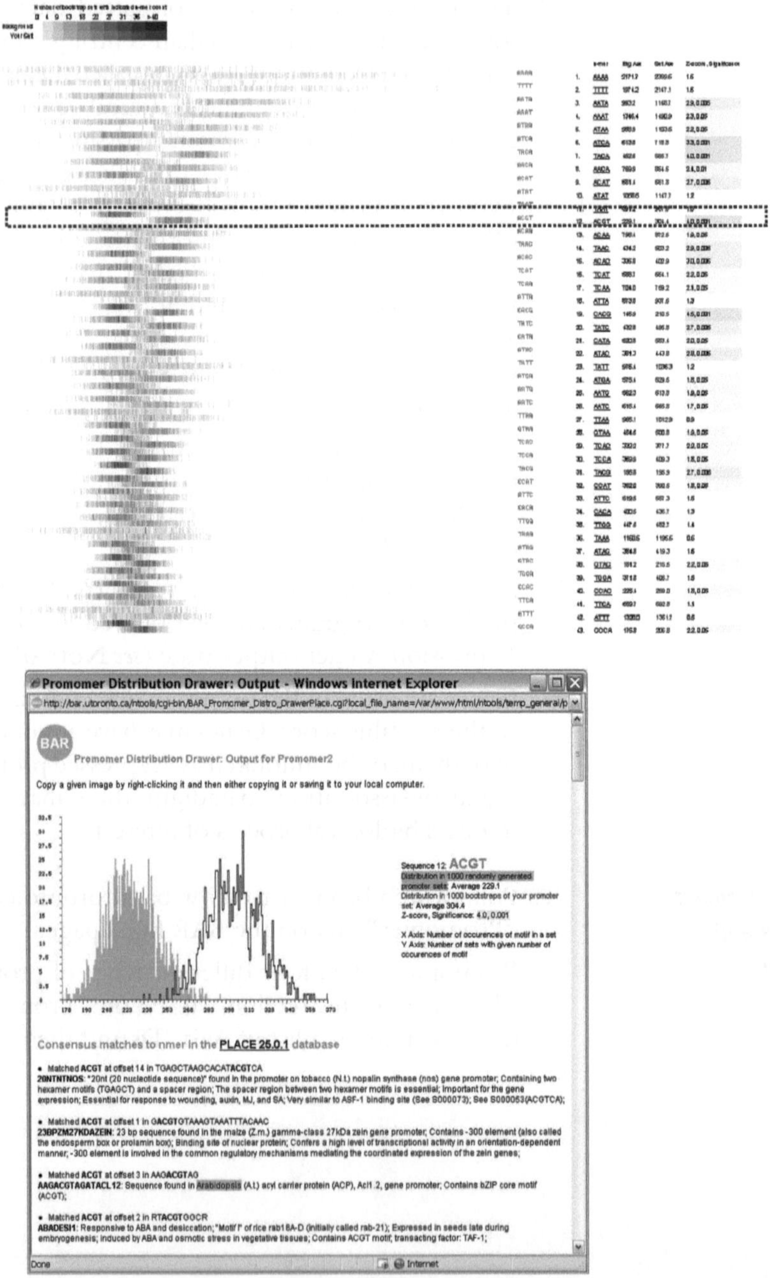

Fig. 3.5. (**A**) An example Promomer 2 output using a test set of 99 genes that are most strongly upregulated by ABA in the mesophyll cells of Arabidopsis plants *(24)* and searching for 4-mer motifs in at least 75 percent of these genes. In each pair of distributions for each *n*-mer, the *top* (*blue*) distribution shown represents the number of sets (out of 1000 randomly generated promoter sets from the Arabidopsis genome having the same number of promoters as the input set) having the indicated occurrences of the given *n*-mer. The *bottom* (*coloured red on a computer screen*) distributions shown on the graph represents the number of sets (out of 1000 randomly generated promoter sets from the input set, with repeats allowed) having the indicated number of instances of the given *n*-mer for each of the *n*-mer identified. The intensity of the colour indicates the number of occurrences according to the colour scale shown. Significantly over-represented words are highlighted in *yellow* in the

count analysis on their promoters (*see* **Note 7** for an overview of the other Promomer types). What this means is that the promoters are broken down into "words" of a user-specified length, and then the number of occurrences of each word are counted. Statistical significance for a given word is established by comparing its frequency in the promoters of the input set AGI IDs to that of random set of promoters.

3. On the Promomer input page, select the data set you would like to search in. There are two upstream data sets of 500 and 1,000 bp promoters, and two downstream data sets, also of 500 and 1,000 bp. The latter sets may be useful for identifying regulatory elements in the 3' UTRs of genes that are coregulated, while the more typical application is to search in the promoters to discover potential regulatory motifs.
4. Use the radio button to select the Promomer 2 option "Identify a statistically over-represented element in a group of genes".
5. Paste in a list of AGI IDs, one per line. These may be copied from Excel or from a text-editing program.
6. Select the number of base pairs the element should contain. This may range from 4–10 bp.
7. Select the minimum percentage of genes in which the identified element should occur (*see* **Note 8**).
8. Click Submit Query and let Promomer run for about 30 s.
9. On the output page, there are two links to results generated by Promomer, one to "View graphical interpretation of results" and one to "Results in plain-text format". Click these links to get the results.
10. The graphical view generates a graph as illustrated in **Fig. 3.5**A.
11. Clicking on a given 4-mer will open another window that will show the distribution of that 4-mer in the 1,000 bootstraps of the input promoter set and in 1,000 random promoters set as more familiar distribution histogram. For example, clicking on ACGT in the above overview will yield a page as illustrated in **Fig. 3.5B** (*see* **Note 9** for tips on saving these images).

Fig. 3.5. (continued) *rightmost column*. The ACGT *n*-mer, surrounded by the *dotted box*, is one of these. Click on this *n*-mer will call up the histogram view. (**B**) An example of the histogram view of a specific Promomer *n*-mer "word". It is clear that ACGT, the core of the ABRE, is over-represented in the input promoters (distribution on the *right*) versus the randomly selected promoters (distribution on the *left*). Also, highlighted in the text below the histogram is the match to any element that has been stored in PLACE *(25)*.

4. Notes

1. There are several other interpretive modes of the eFP Browser beyond looking at the raw values of a gene across different samples. One is to look at the "Relative" values of the gene by switching the "Mode" pull down menu option from Absolute to Relative. The relative mode uses the median expression level or the expression values for control samples as the baseline and then compares the expression level in each sample to this. These values (fold-change above or below median or relative to the control sample) are then log transformed to aid in visualising differences in signal intensity between samples. Another option is to change the "Mode" pull down to "Compare", at which point the user can enter a second AGI identifier in the "Secondary AGI ID" field at the top of the screen. The eFP Browser will then display the comparative log transformed values of the two genes in the eFP output format, relative to their median or control values. This is shown in **Fig. 3.6.**

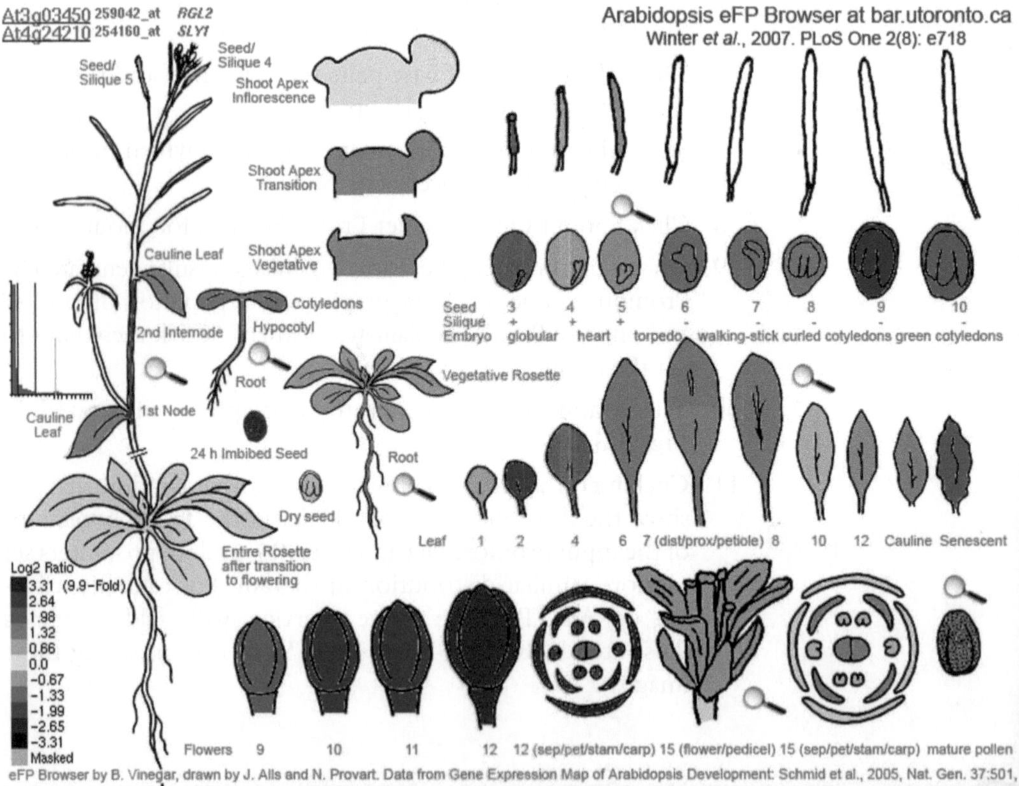

Fig. 3.6. A comparison between the gibberellin response inhibitor protein *RGL2* (At3g03450) and the F-box protein *SLY* (At4g24210). SLY interacts with RGL2 to facilitate its degradation in the presence of GA *(26)*. Darker shading here indicates that one of these two genes dominates – on a computer screen red would indicate that RGL2 dominates, and blue would indicate that SLY dominates – while lighter shading (in yellow on a computer screen) indicates an equal relative level of both, here in the inflorescence shoot apex.

2. Caution in interpreting clutering data is required given that some genes are more highly expressed than others. If the third output option "Average of replicate treatments relative to average of appropriate control" has been selected, the clustering analysis is performed by examining ratiometric fold-change across the different treatments. Currently, there is no filtering for genes with a low level of expression, including those below the suggested background level of 20 raw intensity units. The specific signal intensities for different genes in the AGI identifier list may be checked by placing the AGI IDs in the eFP Browser and reading the colour scale.
3. Note that it is possible to quickly call up the eFP Browser when viewing the heatmap output of Expression Browser by holding down the "e" key on the keyboard and clicking on a given row in the heatmap output of the Expression Browser or simply by clicking on the AGI ID itself in the heatmap. Alternately, one can input the AGI IDs into Expression Browser in the same order as in the log-transformed output and then select "Average of replicate treatments" output option. Use the "unclustered results" link on the output page to get a heatmap that displays the raw signal intensities in the same order as the log-transformed clustered results.
4. Expression Angler offers several additional features. It is possible to drop certain data sets for the calculation of the correlation coefficient, using the Subselect feature: use the link to the "Subselect and Custom Bait Page" on the Expression Angler input page, select the data set in which you wish to select only certain samples for the co-expression analysis, and on the following page, use the checkboxes to select these. In the example shown in **Fig. 3.7B**, all samples except the seed samples were selected for the analysis, in contrast to the analysis illustrated in **Fig. 3.7A**, where seed samples were included.

The above example shows that it is necessary to reconcile our thinking to the idea of networks of genes – some highly connected and others not so – being involved in specific responses and in the development and manifestation of different tissues. **Figure 3.8** shows *r*-value calculations between 19,979 genes whose protein products are predicted to interact, based on orthology with proteins that interact in other organisms. Matt Geisler and Jane Geisler-Lee at Southern Illinois University performed this interolog analysis, and the predictions are available in the Arabidopsis Interactions Viewer, also available at www.bar.utoronto.ca *(19)*. In the two graphs of **Fig. 3.8**, for a given interolog pair, the *r*-value from Abiotic Stress compendium *(5)* is plotted along one axis and its corresponding *r*-value in the a large compendium of 945 AtGenExpress data sets, in the left graph, or against the Developmental Map compendium *(4)*, in the right graph, along the other. Clearly,

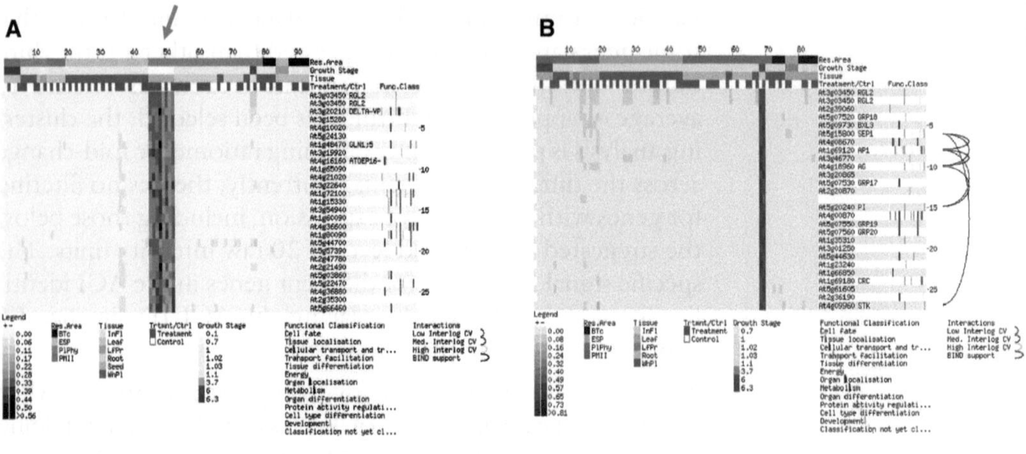

Fig. 3.7. Co-expression analysis with *RGL2* in the Bio-Array Resource Database with (**A**) and without (**B**) seed samples, which are samples 45–52 in the (**A**) panel, pointed to by the *arrow*. Floral homeotic genes, connected by interaction loops on the right in the *bottom panel*, are apparent only in the latter case as *RGL2*'s expression in seeds dominates otherwise, leading to many general seed-specific genes being returned.

the co-expression results are different in one set versus the results from the merged compendium of 945 samples, as they are different from the results from, e.g. the Developmental Map, as indicated by the very broad distribution of the points. This is somewhat in contrast to the statement in *(13)* that says that co-expression results do not change if additional sets are added for the analysis. It is far more appropriate to begin to think of which sets of genes are being expressed in each tissue and how these respond to various stress conditions, i.e. to begin to think of plant biology in a systems biological context *(20, 21)*.

5. It is possible to use the custom bait feature to identify genes that are expressed only in certain samples – even if one does not have a candidate gene in mind. There is a link to the "Subselect and

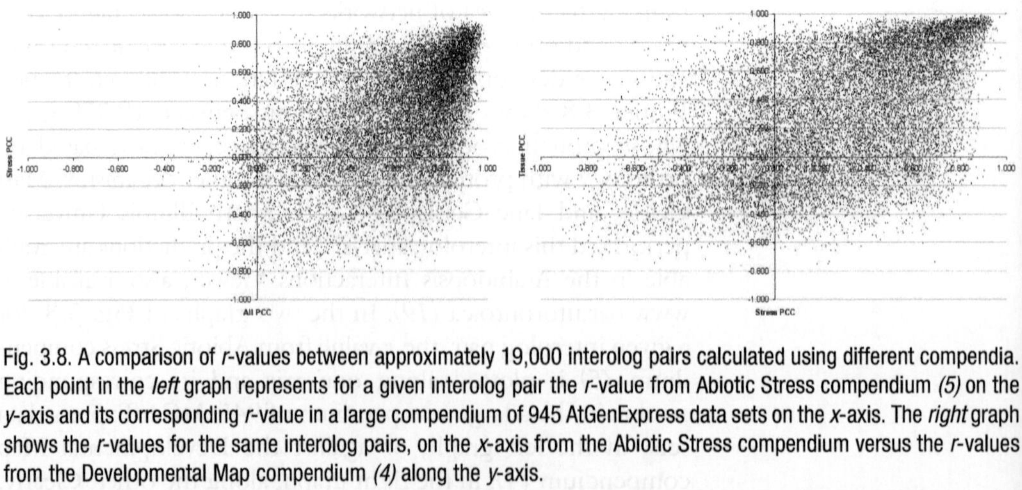

Fig. 3.8. A comparison of *r*-values between approximately 19,000 interolog pairs calculated using different compendia. Each point in the *left* graph represents for a given interolog pair the *r*-value from Abiotic Stress compendium *(5)* on the *y*-axis and its corresponding *r*-value in a large compendium of 945 AtGenExpress data sets on the *x*-axis. The *right* graph shows the *r*-values for the same interolog pairs, on the *x*-axis from the Abiotic Stress compendium versus the *r*-values from the Developmental Map compendium *(4)* along the *y*-axis.

A

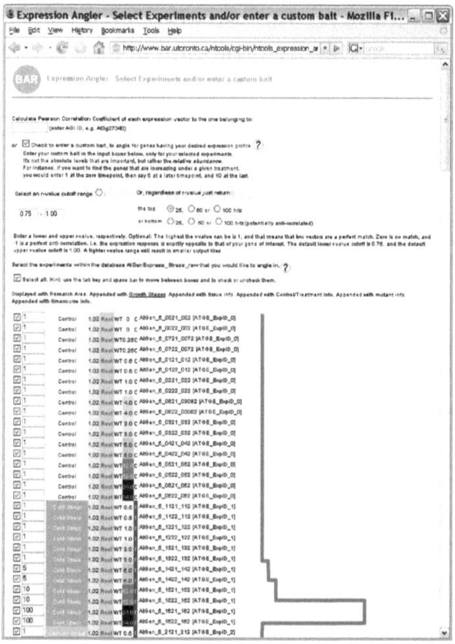

B

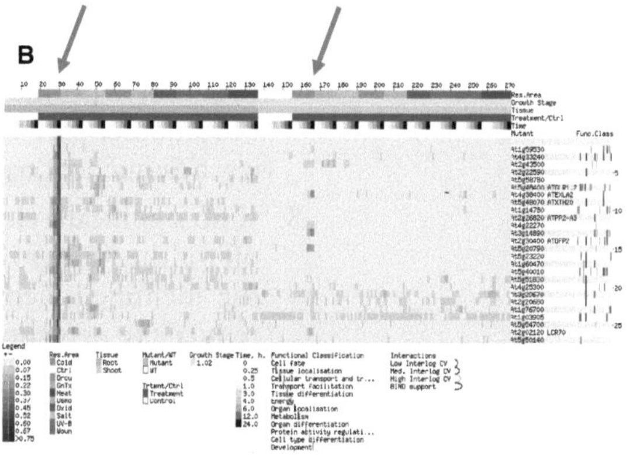

Fig. 3.9. (**A**) Using the Custom Bait feature of Expression Angler to identify genes that are upregulated after 24 h of cold stress in the roots of Arabidopsis plants. The line on the *right* has been added to denote the "shape" of the vector that will be used to search the AtGenExpress Stress data set. The shape of the vector is set by entering values in the fields to the *left* of each sample listed. A value of some arbitrarily high amount, 100 in this case as compared to the background level of 1, is sufficient to identify specific genes. (**B**) Heatmap output of Expression Angler using the "Custom Bait" feature. Genes that are specifically upregulated in roots of plants subjected to cold stress for 24 h are identified, as seen by the darker stripe corresponding to samples 29 and 30, pointed to by the *left arrow*. Note that there is no upregulation in the shoot samples under the same treatment, which are samples 165 and 166 in the heatmap, highlighted by the *rightmost arrow*.

Custom Bait Page" on the Expression Angler input page. An example of the custom bait feature is illustrated in **Fig. 3.9**. The "Check to enter a custom bait" checkbox should be checked, and then arbitrarily high values are entered into input boxes to the left of each sample description listed at the bottom of the webpage. The value that is entered into each box is not so important; rather it is more the shape of the vector that is being designed that is critical. Typically setting the value for a sample to 100 and leaving all other values at 1 will result in the identification of sets of genes that are specific for the given treatment or tissue represented by that sample, insofar as they exist.

6. Expression Angler also provides a link on the output page to a useful program, called Athena, which will analyse the promoters of co-expressed genes for over-representation of known *cis*-elements. A description of the Athena program is available elsewhere *(22)*.

7. If one is interested in the promoter of a particular gene, it is possible to analyse just one promoter at a time with the

Promomer 1 algorithm for over-represented "words". The algorithm works in a similar way to Promomer 2 except that just a single promoter is being analysed. Promomer 3 may be used to identify promoters in the genome that contain an element of interest, while Promomer 4 allows one to search in a subset of promoters for a given motif. If the motif specified in Promomer 3 or 4 contains a "wobble", then it is possible to view a SequenceLogo *(23)* of the matches identified.

8. If no significant elements are identified with a particular cut-off for the "Enter the minimum percentage of genes in which the identified element should occur" field, it may help to reduce the percentage setting.
9. In order to save the output images generated by Promomer or any of the other programs described here, position the mouse pointer over the desired image and right click (click and hold if you are using a Mac) and choose the "Save picture as..." option in the list that appears.

References

1. Lee, H. K., Hsu, A. K., Sajdak, J., Qin, J., and Pavlidis, P. (2004). Coexpression analysis of human genes across many microarray data sets. *Genome Res 14*, 1085–94.
2. Wolfe, C., Kohane, I., and Butte, A. (2005). Systematic survey reveals general applicability of "guilt-by-association" within gene coexpression networks. *BMC Bioinformatics 6*, 227.
3. Zhang, W., Morris, Q., Chang, R., Shai, O., Bakowski, M., Mitsakakis, N., Mohammad, N., Robinson, M., Zirngibl, R., Somogyi, E., Laurin, N., Eftekharpour, E., Sat, E., Grigull, J., Pan, Q., Peng, W.-T., Krogan, N., Greenblatt, J., Fehlings, M., van der Kooy, D., Aubin, J., Bruneau, B., Rossant, J., Blencowe, B., Frey, B., and Hughes, T. (2004). The functional landscape of mouse gene expression. *J Biol 3*, 21.
4. Schmid, M., Davison, T. S., Henz, S. R., Pape, U. J., Demar, M., Vingron, M., Scholkopf, B., Weigel, D., and Lohmann, J. U. (2005). A gene expression map of Arabidopsis thaliana development. *Nat Genet* 37, 501–06.
5. Kilian, J., Whitehead, D., Horak, J., Wanke, D., Weinl, S., Batistic, O., D'Angelo, C., Bornberg-Bauer, E., Kudla, J., and Harter, K. (2007). The AtGenExpress global stress expression data set: protocols, evaluation and model data analysis of UV-B light, drought and cold stress responses. *Plant J 50*, 347–63.
6. Edgar, R., Domrachev, M., and Lash, A. E. (2002). Gene Expression Omnibus: NCBI gene expression and hybridization array data repository. *Nucl Acids Res 30*, 207–10.
7. Rocca-Serra, P., Brazma, A., Parkinson, H., Sarkans, U., Shojatalab, M., Contrino, S., Vilo, J., Abeygunawardena, N., Mukherjee, G., Holloway, E., Kapushesky, M., Kemmeren, P., Lara, G. G., Oezcimen, A., and Sansone, S. A. (2003). ArrayExpress: a public database of gene expression data at EBI. *Current Res Biol 326*, 1075–78.
8. Garcia-Hernandez, M., Berardini, T. Z., Chen, G., Crist, D., Doyle, A., Huala, E., Knee, E., Lambrecht, M., Miller, N., Mueller, L. A., Mundodi, S., Reiser, L., Rhee, S. Y., Scholl, R., Tacklind, J., Weems, D. C., Wu, Y., Xu, I., Yoo, D., Yoon, J., and Zhang, P. (2002). TAIR: a resource for integrated Arabidopsis data. *Funct Integr Genomics 2*, 239–53.
9. Craigon, D. J., James, N., Okyrere, J., Higgins, J., Jotham, J., and May, S. (2004). NASCArrays: a repository for microarray data generated by NASC's transciptomics service. *Nucl Acids Res 32*, D575–D77.

10. Zimmermann, P., Hirsch-Hoffman, M., Hennig, L., and Gruissem, W. (2004). GENEVESTIGATOR. Arabidopsis microarray database and analysis toolbox. *Plant Physiol 136*, 2621–32.
11. Toufighi, K., Brady, M., Austin, R., Ly, E., and Provart, N. (2005). The botany array resource: e-Northerns, expression angling, and promoter analyses. *Plant J 43*, 153–63.
12. Obayashi, T., Kinoshita, K., Nakai, K., Shibaoka, M., Hayashi, S., Saeki, M., Shibata, D., Saito, K., and Ohta, H. (2007). ATTED-II: a database of co-expressed genes and cis elements for identifying co-regulated gene groups in Arabidopsis. *Nucl Acids Res 35*, D863–69.
13. Manfield, I. W., Jen, C.-H., Pinney, J. W., Michalopoulos, I., Bradford, J. R., Gilmartin, P. M., and Westhead, D. R. (2006). Arabidopsis Co-expression Tool (ACT): web server tools for microarray-based gene expression analysis. *Nucl Acids Res 34*, W504–09.
14. Steinhauser, D., Usadel, B., Luedemann, A., Thimm, O., and Kopka, J. (2004). CSB.DB: a comprehensive systems-biology database. *Bioinformatics 20*, 3647–51.
15. Aoki, K., Ogata, Y., and Shibata, D. (2007). Approaches for extracting practical information from gene co-expression networks in plant biology. *Plant Cell Physiol 48*, 381–90.
16. Wille, A., Zimmermann, P., Vranova, E., Furholz, A., Laule, O., Bleuler, S., Hennig, L., Prelic, A., von Rohr, P., Thiele, L., Zitzler, E., Gruissem, W., and Buhlmann, P. (2004). Sparse graphical Gaussian modeling of the isoprenoid gene network in Arabidopsis thaliana. *Genome Biol 5*, R92.
17. Ehlting, J., Provart, N. J., and Werck-Reichhart, D. (2006). Functional annotation of the Arabidopsis P450 superfamily based on large-scale co-expression analysis. *Biochem Soc Trans 34*, 1192–98.
18. Winter, D., Vinegar, B., Nahal, H., Ammar, R., Wilson, G. V., and Provart, N. J. (2007). An 'electronic fluorescent pictograph' browser for exploring and analyzing large-scale biological data sets. *PLoS One 2*, e718.
19. Geisler-Lee, J., O'Toole, N., Ammar, R., Provart, N. J., Millar, A. H., and Geisler, M. (2007). A predicted interactome for Arabidopsis. *Plant Physiol 145*, 317–29.
20. Trewavas, A. (2006). A brief history of systems biology: "Every object that biology studies is a system of systems." *Plant Cell 18*, 2420–30.
21. Gutierrez, R. A., Shasha, D. E., and Coruzzi, G. M. (2005). Systems biology for the virtual plant. *Plant Physiol 138*, 550–54.
22. O'Connor, T. R., Dyreson, C., and Wyrick, J. J. (2005). Athena: a resource for rapid visualization and systematic analysis of Arabidopsis promoter sequences. *Bioinformatics 21*, 4411–13.
23. Crooks, G. E., Hon, G., Chandonia, J.-M., and Brenner, S. E. (2004). WebLogo: a sequence logo generator. *Genome Res 14*, 1188–90.
24. Leonhardt, N., Kwak, J. M., Robert, N., Waner, D. L. G., and Schroeder, J. L. (2004). Microarray expression analyses of Arabidopsis guard cells and isolation of a recessive ABA hypersensitive protein phosphatase 2C mutant. *Plant Cell 16*, 596–615.
25. Higo, K., Ugawa, Y., Iwamoto, M., and Higo, H. (1998). PLACE: a database of plant cis-acting regulatory DNA elements. *Nucl Acids Res 26*, 358–59.
26. McGinnis, K. M., Thomas, S. G., Soule, J. D., Strader, L. C., Zale, J. M., Sun, T.-p., and Steber, C. M. (2003). The Arabidopsis SLEEPY1 gene encodes a putative F-Box subunit of an SCF E3 ubiquitin ligase. *Plant Cell 15*, 1120–30.

Chapter 4

Measurement of Plant Hormones by Liquid Chromatography–Mass Spectrometry

Stacey J. Owen and Suzanne R. Abrams

Abstract

Abscisic acid (ABA) plays a number of key roles in the growth, development, and stress response of plants. For example, it is vital to a plant's response to drought stress, and is the signalling molecule responsible for closure of the stomata in order to promote water conservation. The hormone is rapidly turned over in plant tissue, mainly by oxidation or conjugation. Accurate and sensitive quantification of ABA and its metabolites has made a significant contribution to the knowledge of the role of this hormone, and also of its relationship to the induction of numerous ABA-induced genes in plants. High-performance liquid chromatography coupled with tandem mass spectrometry (LC-MS/MS) has become an essential technique for the analysis and quantification of these compounds.

Key words: HPLC, tandem mass spectrometry, quantification, plant hormones, abscisic acid, abscisic acid metabolites.

1. Introduction

Plant hormones are low molecular weight molecules that are synthesised by a plant and they signal a response to internal or external cues. In performing this signalling role, hormones work to control and direct growth, development, and response to stress. They are present and are biologically active at very low concentrations, from pg/g to ng/g dry weight. Analysis techniques have to be extremely sensitive to be able to observe and quantify hormones at these trace levels. HPLC coupled to tandem mass spectrometry has become the method of choice for the analysis of many plant hormones *(1, 2)*. It requires no derivatisation steps and limited sample preparation, boasts a very high

Sean Cutler, Dario Bonetta (eds.), *Plant Hormones: Methods and Protocols, Second Edition, vol. 495*

DOI 10.1007/978-1-59745-477-3_4 Springerprotocols.com

sensitivity in multiple-reaction monitoring (MRM) mode, and is highly versatile in being able to analyse multiple hormones from a variety of classes in a single run. Quantification of hormones is achieved by the addition of specifically deuterated analogs of the analytes as internal standards.

Abscisic acid (ABA) is a plant hormone that fills a variety of signalling roles in a plant. It is involved in functions, such as stomatal closure, inhibition of shoot growth, induction of storage protein synthesis in seeds, induction and maintenance of dormancy in seeds, and wounding response *(3)*. Metabolism of ABA in plants can take a number of different routes including oxidation, conjugation, or reduction *(4, 5)*. The pool size of ABA at any given moment in plant tissue is a product of transport, synthesis, and metabolism, each of which might be occurring at different rates (resulting in accumulation or loss of ABA over time) or at the same rate (resulting in a steady state concentration of ABA over time). Thus, it is very important to quantify both ABA and its downstream metabolites in order to obtain an accurate picture of the hormone status in a plant tissue. A large number of ABA metabolites have been identified and characterised from various plant tissues *(4)*. Five of these metabolites are quantified, along with ABA (**1.1**, *see* **Fig. 4.1** for all structures) in the current method, including abscisic acid

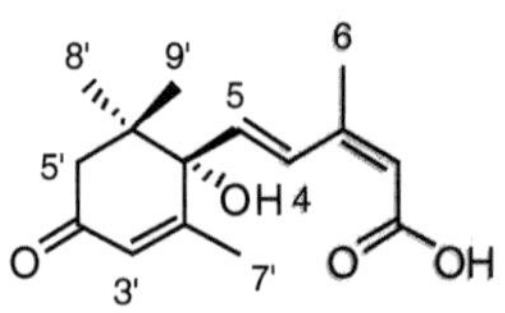

1.1 (+)-*S*-ABA ($C_{15}H_{20}O_4$)

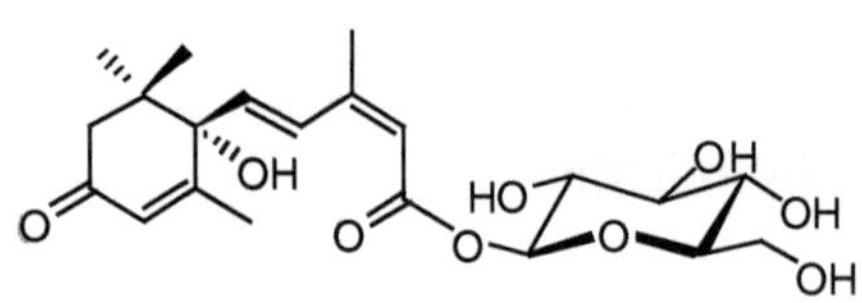

1.2 ABA-GE ($C_{21}H_{30}O_9$)

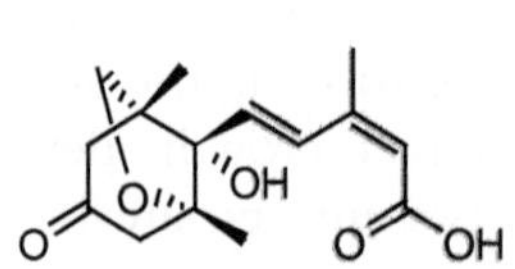

1.3 PA ($C_{15}H_{20}O_5$)

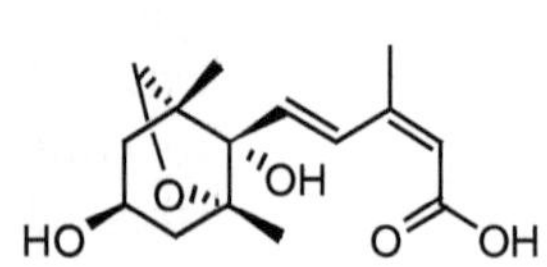

1.4 DPA ($C_{15}H_{22}O_5$)

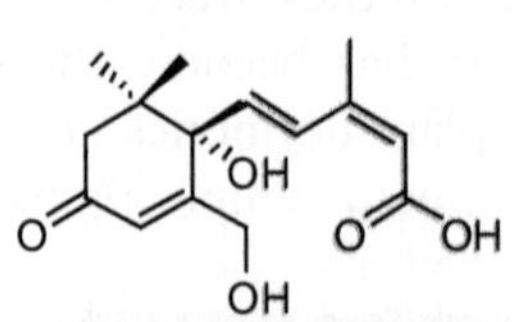

1.5 7'-OH ABA ($C_{15}H_{20}O_5$)

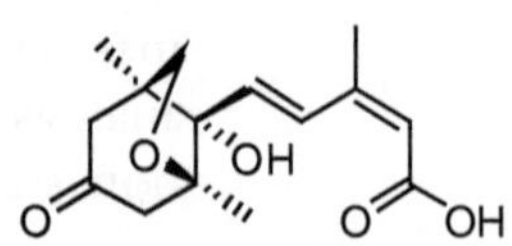

1.6 neoPA ($C_{15}H_{20}O_5$)

Fig. 4.1. Structures of ABA and the five metabolites which are quantified in the current method.

glucose ester (ABA-GE, **1.2**), phaseic acid (PA, **1.3**), dihydrophaseic acid (DPA, **1.4**), 7'-hydroxy abscisic acid (7'-OH ABA, **1.5**), and *neo*phaseic acid (*neo*PA, **1.6**).

2. Materials

Solvents utilised in this procedure, such as acetonitrile, isopropanol, and methanol, are of HPLC grade or better. Water used is de-ionised using a Milli-Q system (18 megaohm). All chemicals and solvents are used as received, without further purification. The acetic acid (glacial) used in the procedure is of 99.99+ % purity.

2.1. Plant Hormone Extraction and Cleanup

1. *Analyte Solution*: This solution contains each of the following analytes at 200 pg/μL in acetonitrile:water:acetic acid, 49.5:50.0:0.5 (v/v/v, *see* **Note 1**): (−)-DPA, (+)-ABA-GE, (−)-PA, (±)-7′-OH ABA, (+)-*neo*PA, and (±)-ABA (all except (±)-ABA (Sigma-Aldrich Canada, Oakville, Ontario, Canada) were synthesised at the National Research Council of Canada, Plant Biotechnology Institute (NRC-PBI), Saskatoon, Saskatchewan, Canada).
2. *Internal Standard Solution*: This solution contains each of the following deuterated versions of the analytes at a concentration of 200 pg/μL in acetonitrile:water:acetic acid, 49.5:50.0:0.5 (v/v/v, *see* **Note 1**): (−)-7′,7′,7′-d_3-DPA, (+)-4,5,8′,8′,8′-d_5-ABA-GE, (−)-7′,7′,7′-d_3-PA, (−)-5,8′,8′,8′-d_4-7′-OH ABA, (−)-8′,8′,8′-d_3-neoPA, and (−)-5,8′,8′,8′-d_4-ABA (all were synthesised at NRC-PBI, Saskatoon, Saskatchewan, Canada).
3. *Extraction Solvent*: isopropanol:water:acetic acid, 80:19:1 (v/v/v).
4. Mixed-mode cation exchange solid-phase extraction cartridges (Oasis MCX SPE, 3 cc, 60 mg, Waters Corporation, Mississauga, Ontario, Canada).
5. *Acidified Water*: water:acetic acid, 99:1 (v/v).
6. *Acidified Methanol*: methanol:acetic acid, 99:1 (v/v).
7. Vacuum manifold for solid-phase extraction.
8. Hydrophilic-lipophilic balance solid-phase extraction cartridges (Oasis HLB SPE, 1 cc, 30 mg, Waters Corporation, Mississauga, Ontario, Canada).
9. *HLB Elution Solvent*: water:acetonitrile:acetic acid, 69:30:1 (v/v/v).

10. *Reconstitution Solution:* This solution contains each of the following deuterated versions of the analytes at 100 pg/μL in acetonitrile: water, 15:85 (v/v) containing 0.1% acetic acid (*see* **Note 1** and **Note 2**): (+)-4,5-d_2-ABA-GE and (±)-3′,5′,5′,7′,7′,7′-d_6-ABA (both were synthesised at NRC-PBI, Saskatoon, Saskatchewan, Canada, *see* **Note 3**).

2.2. LC-MS/MS Analysis and Quantification

1. An HPLC system coupled to a triple quadrupole mass spectrometer via an electrospray source is required. The HPLC must be capable of delivering a gradient of at least two solvents. The mass spectrometer must be capable of acquiring highly sensitive negative ion, multiple-reaction monitoring (MRM) data.
2. Genesis C18 reversed-phase column (2.1 × 100 mm, 4 μm pore size, Chromatographic Specialties, Brockville, Ontario, Canada) with an Opti-Guard C18 guard column (1 mm, Optimize Technologies, Inc., Oregon City, OR, USA).
3. *Calibration Solutions*: Six to eight calibration solutions should be prepared in order to adequately describe the calibration equations for each analyte. These solutions should contain each analyte in equal concentration in any particular solution, and varying concentrations of the analytes between solutions, over the range of concentrations expected in samples (for example, 2–1,000 pg/μL of each of the six analytes, *see* **Note 4**), along with 100 pg/μL of each of the six internal standard compounds. Generally, these solutions are prepared in the same manner as samples, in that the analytes and internal standards are added together in solution, this solution is completely evaporated to dryness under vacuum, and then it is reconstituted in *Reconstitution Solution* to arrive at the final concentration desired.

3. Methods

This method is applicable to a variety of plant tissue types, such as leaves, roots, stems/branches, seeds, berries, and flowers, and requires a minimum of 50 mg dry weight of plant tissue. The extraction procedure takes several days to complete and the LC-MS/MS analysis takes approximately 30 min per sample. However, due to the number of analytes and the complexity of the procedure, initial setup and optimisation of the LC-MS/MS system can take a number of weeks to accomplish, and should be performed by experienced personnel.

3.1. Plant Hormone Extraction and Cleanup

1. Collect plant tissue for analysis and immediately freeze the tissue by immersing it in liquid nitrogen (*see* **Note 5**). Lyophilise the tissue in a freeze-drier and then store it in a –20°C freezer until extraction (*see* **Note 6**).
2. Grind lyophilised plant tissue to a fine and homogeneous powder in a ball mill (e.g. Mini-BeadBeater-96, Biospec Products, Inc., Bartlesville, Oklahoma, USA, *see* **Note 7**).
3. Weigh out 50 mg of homogenised tissue into a 15 mL Falcon tube. A separate 100 μL portion of *Analyte Solution* should also be taken through the extraction process alongside samples periodically for quality control (QC) purposes and to double check recovery. To each sample and QC, add 100 μL of the *Internal Standard Solution* and 3 mL of *Extraction Solvent.* Vortex mix each solution for one minute before placing the tubes on an orbital shaker kept in the refrigerator (4°C) at ~350–400 rpm for 18–24 h.
4. After extraction, vortex mix each solution for one minute, and then place them in a centrifuge at ~4,000–5,000 rpm for 10 min. Transfer each supernatant (~3 mL) to an individual disposable glass test tube (13 × 100 mm), or other container suitable for the particular apparatus used to evaporate the solvent. Re-extract the solid material a second time using 0.5 mL of *Extraction Solvent.* Vortex and centrifuge as above, and add this solvent to the disposable glass test tube containing the original extract.
5. Completely evaporate the solvent using, for example, a Büchi Syncore Polyvap (Büchi, Switzerland, *see* **Note 8**). On the Polyvap, the settings are 30°C, 340 rpm, and a vacuum program taking the pressure from 130 to 30 mbar over one hour, after which it is held at 30 mbar until dry.
6. Initial (optional) cleanup by mixed-mode cation exchange solid-phase extraction using Oasis MCX SPE cartridges (*see* **Note 9**).
   a. Reconstitute each dried extract/QC in 2 mL of *Acidified Water* and vortex mix until dissolved (*see* **Note 10**).
   b. Prepare each cartridge first with a wash of 3 mL of *Acidified Methanol*, and then equilibrate it using 3 mL of *Acidified Water.* Discard these fractions into waste vials.
   c. Each reconstituted sample/QC should now be added to a separate cartridge and allowed to drip through at a rate of one drop every 1–2 s (*see* **Note 11**). The disposable glass sample vials should then be rinsed with 0.5 mL of *Acidified Water*, and this wash added

to the appropriate MCX SPE cartridge. Discard these fractions into waste vials.

d. Wash the most polar components off of the MCX SPE cartridge using 1 mL of *Acidified Water* at a slow flow rate as in **Step 6c**. Discard this fraction into waste vials.

e. Using new microcentrifuge tubes to collect each purified fraction, elute the analytes from each column using 1.5 mL of *Acidified Methanol* (*see* **Note 12**). Dry these fractions down completely in a centrifugal evaporator.

7. Cleanup of samples/QC by hydrophilic–lipophilic balance solid-phase extraction using Oasis HLB SPE cartridges (*see* **Note 13**).

   a. Reconstitute each dried sample/QC in 100 μL of *Acidified Methanol*, and vortex mix until all components that will dissolve are in solution (*see* **Note 10**).

   b. Add 900 μL of *Acidified Water* to each sample/QC and vortex mix until dissolved (*see* **Note 10**).

   c. Prepare each cartridge first with a wash of 1 mL of *Acidified Methanol*, and then equilibrate it using 1 mL of *Acidified Water*. Discard these fractions into waste vials.

   d. Each reconstituted sample/QC should now be added to a separate cartridge and allowed to drip through at a rate of one drop every 1–2 s (*see* **Note 11**). Each microcentrifuge vial should then be rinsed with 0.5 mL of *Acidified Water*, and this wash added to the appropriate HLB SPE cartridge. Discard these fractions into waste vials.

   e. Elute the most polar components from each HLB SPE cartridge using 1 mL of *Acidified Water* at a slow flow rate as in **Step 7d**. Discard this fraction into waste vials.

   f. Using new microcentrifuge tubes to collect each purified fraction, elute the analytes from each column using 1 mL of *HLB Elution Solvent*. Dry these fractions down completely in a centrifugal evaporator.

8. Prepare a set of non-extracted QCs and blanks along with each batch of samples in order to test pipetting equipment and technique, to check all solutions (*Analyte Solution*, *Internal Standard Solution*, and *Reconstitution Solution*), to test for injection carry-over, and for the purpose of having a standard response to compare to for the determination of recovery in samples (*see* **Note 14**).

9. Reconstitute samples and QCs in 200 μL of *Reconstitution Solution*. This gives a final concentration of 100 pg/μL of each internal standard, 100 pg/μL of the two recovery standards, and, where added, 100 pg/μL of each analyte.

3.2. LC-MS/MS Analysis and Quantification

1. Use a portion of the *Reconstitution Solution* to infuse as a "tuning mix" to tune the triple quadrupole mass spectrometer and determine the optimum negative-ion electrospray voltage, lens/ion guide settings, resolution settings, gas flows, and temperatures for d_2-ABA-GE and d_6-ABA (*see* **Note 15**).
2. Determine the optimum fragmentation settings for each of the six analyte/IS pairs in the method (*see* **Table 4.1** for mass-to-charge (m/z) MRM transition mass-to-charge ratios for each ion). This includes the more general aspect of finding a good compromise for the collision cell pressure that will work well for all of the ions, as this cannot be changed during a run. It also includes parameters that can be optimised and set for

Table 4.1
MRM transitions and conditions used on the Quattro Ultima triple quadrupole mass spectrometer. CV = cone voltage (in volts, V), CE = collision energy (in electron volts, eV), DT = dwell time (in seconds, s), and RW = retention window (in minutes, min). The inter-channel time is set to 0.04 s, while the inter-scan time is set to 0.1 s

Analyte	Precursor m/z	Product m/z	CV (V)	CE (eV)	DT (s)	RW (min)
DPA	281	171	30	16	0.4	0.0–7.0
d_3-DPA	284	174				
ABA-GE	425	263	40	9	0.3	7.0–9.5
d_5-ABA-GE	430	268				
d_2-ABA-GE	427	265				
PA	279	139	20	11	0.3	8.5–10.4
d_3-PA	282	142				
7'-OH ABA	279	151	20	14	0.4	9.9–11.4
d_4-7'-OH ABA	283	154				
neoPA	279	205	40	11	0.4	11.0–12.5
d_3-neoPA	282	208				
ABA	263	153	20	9	0.4	12.5–15.5
d_4-ABA	267	156				
d_6-ABA	269	159				

each particular ion, such as collision energy and cluster reduction mechanisms (e.g. cone voltage, *see* **Note 16**).

3. Using a Genesis C18 reversed-phase column with an OptiGuard C18 guard column, the HPLC parameters are: column temperature = 34°C, injection volume = 10 μL, total run time = 30 min, and a solvent gradient program and flow rate as shown in **Table 4.2** (*see* **Note 17**). Determine the retention times of each analyte/IS pair and then set the retention windows in which the MRM signals are to be collected in the data collection program (*see* **Note 18**).
4. Once all LC-MS/MS parameters have been determined, analyse all *Calibration Solutions* and construct calibration curves for each analyte/IS pair based on the relative response of the analyte to the IS (*see* **Notes 4** and **19**). Also construct response factor (RF) calibration curves for d_5-ABA-GE, using d_2-ABA-GE as internal standard, and for d_4-ABA, using d_6-ABA as internal standard (*see* **Note 20**). These curves are used to determine recovery of analytes from the plant tissue (*see* **Note 21**).

Table 4.2
HPLC gradient conditions used on the Waters 2695 quaternary HPLC system. Solvent A = acetonitrile, Solvent B = milli-Q water, and Solvent C = 5% aqueous acetic acid. The Curve column indicates the rate at which the solvent changes proportions. Each number is associated with a different curve shape for that change (6 being linear)

Time (min)	%A	%B	%C	Curve	Flow (mL/min)
0.0	15.0	84.2	0.8	1	0.20
10.0	35.0	64.2	0.8	6	0.20
14.2	40.0	59.2	0.8	10	0.20
15.0	60.0	39.2	0.8	6	0.20
18.2	60.0	39.2	0.8	6	0.20
18.3	99.2	0.0	0.8	6	0.20
18.4	99.2	0.0	0.8	1	0.35
19.5	99.2	0.0	0.8	1	0.35
20.5	15.0	84.2	0.8	6	0.35
21.0	15.0	84.2	0.8	1	0.20
30.0	15.0	84.2	0.8	1	0.20

5. Analyse samples, QCs, and blanks prepared in **Section 3.1, Step 9**, and quantify the six analytes in each solution using the calibration curves prepared above.

4. Notes

1. This solution should be stable under acidic conditions for at least six months. It should be stored in the dark due to the light sensitivity of racemic 2*Z*,4*E*-ABA which isomerises to 2*E*,4*E*-ABA upon exposure to light. The ABA used can be racemic, but should not contain mixed 2*Z* and 2*E* isomers.
2. The *Reconstitution Solution* should be prepared in the starting solvent of the gradient used for the HPLC separation.
3. The two deuterated compounds used in the *Reconstitution Solution* are used as recovery standards (calculation of which is explained in **Note 20**).
4. With the exception of PA, which has a quadratic curve, the calibration curves of the other five analytes are generally linear over the concentration range of approximately 2–1,000 pg/μL. Extending the curve beyond this range gives rise to non-linear relationships of concentration and response. Often, a second set of calibration curves covering higher concentrations, for example, from 1,000–5,000 pg/μL, are constructed for samples containing high levels of analytes.
5. Flash-freezing of a sample stops all biological processes from occurring and effectively “freezes” the hormone pools at the concentrations they were at when the sample collection occurred.
6. Lyophilised samples are also stable at room temperature if required, for example, during shipping, as long as they are kept in the dark (*see* **Note 1**)
7. A mortar and pestle may also be used to grind the dried plant tissue; however, in re-using this apparatus for different samples, there is potential for cross-contamination. Use of one disposable, plastic vial per sample, along with stainless steel grinding beads, will effectively eliminate the possibility of cross-contamination. Use of stainless steel beads of at least 3.2 mm diameter is recommended and should be able to homogenise even the toughest dried tissues. The number of beads and shaking time required will depend on tissue type, but, in general, shaking for 1–2 min with five beads is a good starting point.
8. A rotary evaporator may be used; however, it allows the processing of only one sample at a time. A centrifugal concentrator

may also be used, although it may take more time. The Polyvap can evaporate the solvent from up to 96 extracts in about 5–6 h. The disposable glass test tubes called for in **Section 3.1, Step 4** fit perfectly inside the re-useable test tubes provided with the Polyvap, and then can be discarded rather than washed and re-used, which would risk cross-contamination.

9. This step is optional for many types of plant tissues, and was introduced more for the cleanup of berry samples which contain high concentrations of cationic anthocyanins.
10. It may be necessary to sonicate the extract for a short time at this point to help the dissolution process. However, even with sonication, not all components may dissolve.
11. This low flow rate allows time for the analytes to interact with, and bind to, the stationary phase.
12. The majority of the anthocyanins, being cationic, will remain on the cartridge.
13. This step was introduced to wash all of the highly polar compounds, such as salts and sugars, from the extract. These types of compounds are generally present in very high concentrations and contribute significantly to ion suppression in the MS analysis step.
14. It is recommended to make a Reconstitution Solvent blank (requires no dry down step), an Internal Standard Solution blank (100 μL of *Internal Standard Solution*, dried, and reconstituted in 200 μL of *Reconstitution Solution*), and at least one QC solution (for example, 100 μL of *Analyte Solution* plus 100 μL of *Internal Standard Solution*, dried, and reconstituted in 200 μL of *Reconstitution Solution*). These should be injected more than once, at intervals, over the course of a batch of samples.
15. For example, an LC-MS/MS system that has been used for this work is an Alliance 2695 HPLC coupled to a Quattro Ultima triple quadrupole mass spectrometer via a Z-spray electrospray (ES) source (Waters Corporation, Mississauga, Ontario, Canada). Both the HPLC and the MS were controlled by MassLynx v. 4.0 software (Waters Corporation, Mississauga, Ontario, Canada). Typical instrument settings include: capillary potential = 2.75 kV, source temperature = 120°C, desolvation temperature = 350°C, cone gas flow rate = 100 L/h, desolvation gas flow rate = 650 L/h, and quadrupole low and high mass resolutions = 12.0.
16. For example, on the LC-MS system introduced in **Note 15**, the collision cell pressure is set to 1.3×10^{-3} mbar using argon as collision gas and the MRM parameters of collision energy, cone voltage, and dwell time are shown in **Table 4.1**.

One method that can be used to determine the optimum collision energy (CE) and cone voltage (CV, or their equivalents on other instrument types) is to run a series of flow injections for each ion, in which the cone voltage in any one scan is kept constant, while the collision energy is varied. This can be done over a number of cone voltages, for example:

Injection #1 – ABA: all scans: CV = 10 V, scan#1: CE = 5 eV, scan#2: CE = 7 eV, scan#3: CE = 9 eV, scan#4: CE = 11 eV, scan#5: CE = 13 eV.

Injection #2 – ABA: all scans: CV = 20 V, scan#1: CE = 5 eV, scan#2: CE = 7 eV, scan#3: CE = 9 eV, scan#4: CE = 11 eV, scan#5: CE = 13 eV.

Injection #3 – ABA: all scans: CV = 30 V, scan#1: CE = 5 eV, scan#2: CE = 7 eV, scan#3: CE = 9 eV, scan#4: CE = 11 eV, scan#5: CE = 13 eV.

A response curve can then be constructed from this data and the optimum settings can be read from this graph (*see* **Fig. 4.2**).

17. The gradient conditions given in **Table 4.2** have been used on the LC-MS system described in **Note 15**. For other systems which do not allow a nonlinear rate at which the solvent proportions are changed within a step ("Curve"

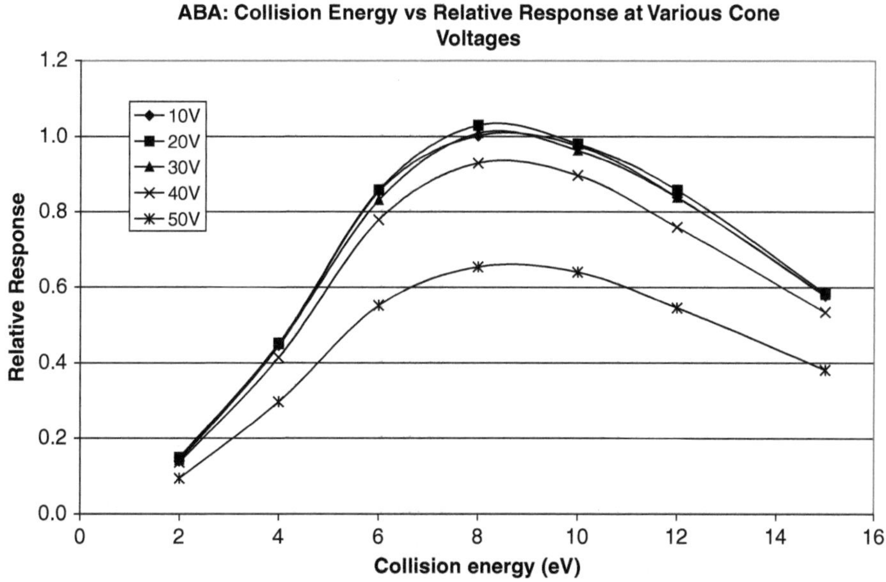

Fig. 4.2. Plot of collision energy (CE, eV) versus relative response of ABA (relative to d_6-ABA) at five different cone voltage (CV, V) settings. From this plot, the optimum CE and CV settings can be determined.

column in **Table 4.2**) some adjustment will be necessary. In doing this adjustment, particular attention must be given to the retention times of ABA-GE and PA, which are the two compounds that have the greatest possibility of overlapping. In fact, using different types of C18 columns or slightly different gradients can cause their elution order to change. Reducing the overlap of these two compounds will minimise the amount of time that MRM signals need to be acquired for both sets of ions at one time, and will therefore increase the duty cycle and sensitivity.

18. Batch to batch variability of Genesis columns has been found to be quite small and only slight adjustments of the retention windows have been necessary when starting to use a new column. An example of the retention windows used on the LC-MS system described in **Note 15** is given in **Table** 4.1.
19. Relative response values can be calculated by the equation:

    Analyte Response = Analyte Area × (IS Concentration / IS Area)
20. Adjustment of the d_2-ABA-GE signal is necessary in order to account for the contribution of the M+2 isotope of d_0-ABA-GE. Theoretically, the isotopic contribution of d_0-ABA-GE to the MRM transition used for d_2-ABA-GE (427 > 265) is 2.1% of the total d_0-ABA-GE MRM signal. This contribution should be subtracted from the d_2-ABA-GE signal, before the response is calculated, for all calibration solutions, standards, and samples.
21. Recovery is determined by calculating the ratio of the d_4-ABA response (using d_6-ABA as IS) in a sample to the average d_4-ABA response (using d_6-ABA as IS) in standards (e.g. IS blanks and QCs), as determined from the RF calibration curve prepared in **Section 3.2, Step 4**.

Acknowledgments

This method has been developed by a team of PBI-NRC researchers: Sue Abrams, Steve Ambrose, Vera Čekić, Sheila Chiwocha, Allan Feurtado, Pearla Gai, Marek Galka, Patrycja Galka, Daiqing Huang, Monika Lafond, Ken Nelson, Stacey Owen, Andrew Ross, Irina Zaharia, and Rong Zhou. The authors thank Irina Zaharia and Andrew Ross for critical reading of the manuscript and acknowledge the National Research Council of Canada's Genomics and Health Initiative (GHI) and Crops for Enhanced Human Health (CEHH) programs for funding. This contribution is NRCC no. 44848.

References

1. Ross, A. R. S., Ambrose, S. J., Cutler, A. J., Feurtado, J. A., Kermode, A. R., Nelson, K., Zhou, R., and Abrams, S. R. (2004) Determination of endogenous and supplied deuterated abscisic acid in plant tissues by high-performance liquid chromatography–electrospray ionization tandem mass spectrometry with multiple reaction monitoring. *Anal Biochem* 329, 324–33.
2. Chiwocha, S. D. S., Abrams, S. R., Ambrose, S. J., Cutler, A. J., Loewen, M., Ross, A. R. S., and Kermode, A. R. (2003) A method for profiling classes of plant hormones and their metabolites using liquid chromatography–electrospray ionization tandem mass spectrometry: an analysis of hormone regulation of thermodormancy of lettuce (*Lactuca sativa* L.) seeds. *Plant J* 35, 405–17.
3. Marion-Poll, A., and Leung, J. (2006) Abscisic acid synthesis, metabolism and signal transduction, in *Plant Hormone Signaling* (Hedden, P., and Thomas, S. G., eds.), Blackwell Publishing, Oxford, UK, pp. 1–35.
4. Zaharia, L. I., Walker-Simmons, M. K., Rodríguez, C. N., and Abrams, S. R. (2005) Chemistry of Abscisic Acid, Abscisic Acid Catabolites and Analogs. *J. Plant Growth Regul.* 24, 274–84.
5. Nambara, E., and Marion-Poll, A. (2005) Abscisic Acid Biosynthesis and Catabolism. *Annu Rev Plant Biol* 56, 165–85.

Chapter 5

Measurement of Abscisic Acid and Gibberellins by Gas Chromatography/Mass Spectrometry

Masanori Okamoto, Atsushi Hanada, Yuji Kamiya, Shinjiro Yamaguchi, and Eiji Nambara

Abstract

Gas chromatography–mass spectrometry (GC-MS)-based analysis is an accurate and sensitive method to quantify plant hormones. This method is commonly used for analysis of low-molecular-weight compounds, such as abscisic acid (ABA), gibberellins (GAs), auxins, and brassinosteroids. Procedures are composed of four major steps: extraction, fractionation, derivatization, and detection. Here, we describe a protocol for quantification of ABA and GAs.

Key words: Abscisic acid, gibberellins, gas chromatography–mass spectrometry (GC-MS), internal standard, derivatization.

1. Introduction

Plant hormones are low-abundant signaling molecules that regulate plant growth and development. Quantification of endogenous hormone levels is a key to understanding their mode of action. Among several methodologies, either mass spectrometry combined with gas chromatography (GC-MS) or with liquid chromatography (LC-MS) is the preferred choice for definitive identification and quantification of plant hormones. Because most plant hormones can be analyzed by either GC- or LC-MS, the choice is dependent on the availability of the instrument. However, it also depends on the chemical nature of target compounds; for example, GC-MS is not suitable for some compounds due to their instability (heat-liability), high polarity, and difficulty in derivatization.

GC-MS-based analysis takes advantage of the high resolution of a capillary column that allows separations of target compounds

Sean Cutler, Dario Bonetta (eds.), *Plant Hormones: Methods and Protocols, Second Edition, vol. 495*

DOI 10.1007/978-1-59745-477-3_5 Springerprotocols.com

from numerous impurities prior to MS analysis. However, multiple fractionation steps are also necessary prior to GC-MS analysis to enrich the target hormones for reliable quantifications because of their low abundance in crude plant extracts. Therefore, the use of internal standards is required to monitor the recovery of target compounds throughout fractionation. The internal standard is ideally a synthesized authentic compound labeled with stable isotope, such as ^{2}H or ^{13}C. Quantification is achieved on GC-MS by comparing the amount of parent or fragment ions from unknown levels of the endogenous compound and from known levels of the added internal standard. Because most plant hormones are not volatile, GC-MS analysis usually requires derivatization of the target compounds to reduce its boiling point. Carboxyl and hydroxyl groups are targets for derivatization by such modifications as methylation and trimethylsilylation. Because a number of specific fragment ions are often generated from a compound upon electron impact (EI) ionization, the full-scan mass spectrum in combination with the retention time on GC are reliable parameters for the identification of known compounds. In practice, selected ion monitoring (SIM), rather than full-scan (total ion) monitoring, is normally used to quantify plant hormones, because the SIM mode provides better sensitivity. In this chapter, we describe a GC-MS-based protocol for measurements of ABA and GAs.

2. Materials

2.1. Extraction

1. 80% (v/v) aqueous methanol (*see* **Note 1**)
2. Isotope-labeled authentic compounds (e.g., [$^{2}H_6$]ABA, [$^{2}H_2$]GA_1 and [$^{2}H_2$]GA_4 (*see* **Note 2**)
3. 2,4-di-*tert*-butyl-4-methylphenol (BHT)
4. 1 *N* HCl
5. *n*-Hexane
6. Ethyl acetate
7. 2% $NaHCO_3$ aqueous solution
8. 10% (v/v) aqueous methanol containing 0.1% acetic acid
9. Sodium sulfate anhydrate

2.2. Column Fractionation and Derivatization

1. Methanol
2. Distilled water
3. Acetic acid

4. Disposable reverse phase cartridge column (Varian, Bond Elut C18, *see* **Note 3**)
5. Insoluble polyvinylpyrrolidone (PVP)
6. 0.1 M phosphate buffer (pH 8.5)
6. Disposable ion exchange cartridge column (Varian, DEA, *see* **Note 3**)
7. High Performance Liquid Chromatograph (HPLC) system (Delta 600 system, Waters)
8. Octadecylsiloxane (ODS) HPLC column (SHISEIDO, CAPCELL PAC C18 5 μm particle size, 6 mm i.d. × 250 mm)
9. Anion exchange HPLC column (Senshu Pak $N(CH_3)_2$, 4.6 mm i.d. × 150 mm).
10. Ethereal diazomethane (*see* **Note 4**)
11. *N*-methyl-*N*-trimethylsilyltrifluoroacetamide (MSTFA)
12. Acetone

2.3. Quantification by GC-MS

1. Gas Chromatograph (Agilent 6870 series GC system)
2. Mass Spectrometer (JMS-Automass SUN, JEOL)
3. Capillary column (DB-1, 250 mm i.d. × 15 m, film thickness of 0.25 mm, J&W Scientific, Folsom, CA)

3. Methods (*see* Fig. 5.1)

3.1. Extraction

1. Plant tissue is harvested, frozen immediately in liquid N_2 and stored at –80°C.
2. To extract the ABA and GAs, the frozen tissue is ground to a powder by using a mortar and pestle with liquid N_2 (*see* **Note 5**). The resultant powder is soaked in 80% methanol containing 0.1 mg/ml 2,4-di-tert-butyl-4-methylphenol (BHT) (*see* **Note 1**). The isotope-labeled internal standards are added prior to extraction (*see* **Note 6**). Each sample is shaken overnight on ice, in darkness, and centrifuged at 3,000 *g* for 10 min at 4°C. The precipitate is re-extracted two times, and the collected supernatant is evaporated under vacuum to eliminate the methanol.
3. The resultant aqueous solution is diluted with distilled water to 5–10 ml and transferred to new glass tube. This aqueous phase is acidified with 1 *N* HCl to adjust the pH to 2–3 (*see* **Note 7**).
4. The ABA and GAs are extracted by partitioning with 10 ml of ethyl acetate, and collected in a new glass tube. This procedure

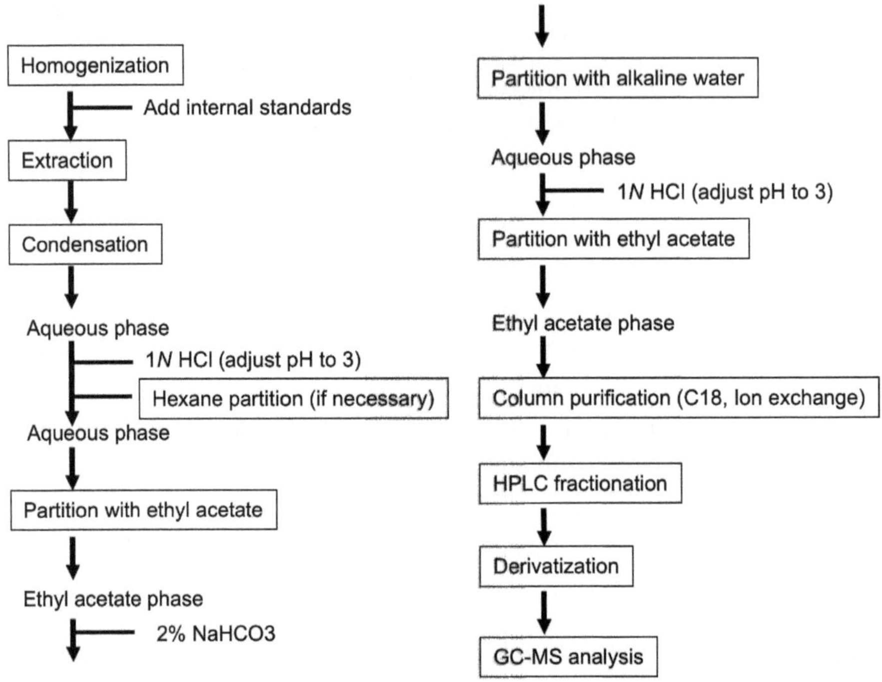

Fig. 5.1 Flowchart for measurements of ABA and GAs. ABA and GAs are weak acids and can be purified through same extraction and purification procedures.

is repeated three times. The ethyl acetate fraction is concentrated to 10 ml under N_2 stream at 30°C (*see* **Note 8**).

5. Ten milliliters of 2% $NaHCO_3$ aqueous solution is added and mixed vigorously (*see* **Note 9**). The alkaline aqueous phase is collected in a new glass tube. This procedure is repeated three times.

6. The aqueous fraction is then adjusted to a pH of 2–3 with 1 *N* HCl. The ABA and GAs are extracted by partitioning with 10 ml of ethyl acetate, and the ethyl acetate fraction is collected in a new glass tube. This procedure is repeated three times. One gram of sodium sulfate anhydrate is added to the ethyl acetate fraction to eliminate residual water. The ethyl acetate fraction is filtered to eliminate sodium sulfate, and dried under vacuum.

3.2. Column Fractionation and Derivatization

1. To start sample fractionation, a reverse-phase cartridge column (Bond Elut C18, Varian) is conditioned with methanol containing 1% (v/v) acetic acid and then with 1% (v/v) aqueous acetic acid (*see* **Notes 7** and **10**). Dried sample is dissolved in 100 μl of 50% (v/v) aqueous methanol and applied to the column. ABA and GAs are eluted with 1 ml of 80% (v/v) aqueous methanol containing 1% (v/v) acetic acid. The eluted

fraction is collected in a new glass tube and dried under vacuum (*see* **Note 7**).

2. Five grams of PVP is suspended and washed three times with distilled water. The PVP is then loaded onto a Bond Elut Reservoir, and is washed with 0.1 M phosphate buffer. The sample from **step 1** is dissolved in 0.1 M phosphate buffer and applied to the PVP column. The ABA and GAs are eluted with 5 ml of 0.1 M phosphate buffer. The eluted fraction is acidified with 1 *N* HCl to pH 2–3 and extracted with ethyl acetate three times. The combined ethyl acetate fractions are dehydrated by adding sodium sulfate anhydrate, filtered, and dried under vacuum (*see* **Note 7**).
3. An ion exchange cartridge column is conditioned with methanol. The dried sample from **step 2** is dissolved in 100 μl of methanol, applied to the column, and washed with methanol. The ABA and GAs are eluted with methanol containing 0.1% (v/v) acetic acid. The volumes for washes and elution depend on column size and are recommended by manufacturer (*see* **Note 7**). The eluate fraction is dried under vacuum.
4. The sample from **step 4** is dissolved in 20 μl of 10% methanol containing 0.1% acetic acid and made up to 200 μl with 0.1% acetic acid. The sample is subjected to reverse-phase HPLC fractionation. The solvents for HPLC are as follows: Solvent A, 0.1% (v/v) aqueous acetic acid; Solvent B, 0.1% (v/v) acetic acid in methanol. Sample separation is carried out by a linear gradient of increasing solvent B (keep 20% solvent B for 5 min and increase 20–80% solvent B over 35 min) at a flow rate of 1.5 ml/min. The column temperature is kept constant at 40°C. UV detection is set to 254 nm. Fractions are collected for GA_1 during 20–22 min, ABA during 26–28 min, and for GA_4 during 36–39 min (*see* **Note 12**). Each fraction is dried under vacuum, dissolved in acetone, and transferred to a small glass tube.
5. After the removal of acetone under vacuum, the ABA and GAs are methylated with fresh ethereal diazomethane. After incubation for 3 min, sample is dried under vacuum. GAs are subsequently trimethylsilylated with MSTFA (20 μl) at 80°C for 30 min and cooled down on ice for 5 min. The derivatized samples are subjected to GC-MS analysis, or stored in the freezer (*see* **Notes 4** and **11**).

3.3. Quantitative Analysis by GC-MS

1. Quantification is performed on a mass spectrometer (JMS-Automass SUN, JEOL) combined with a gas chromatograph (Agilent 6870 series GC system) equipped with a capillary column (DB-1, 0.25 mm i.d. × 15 m, film thickness 0.25 μm). Each sample (1–5 μl) is injected into the GC and quantified by selected ion monitoring MS mode (SIM). Detailed GC-MS

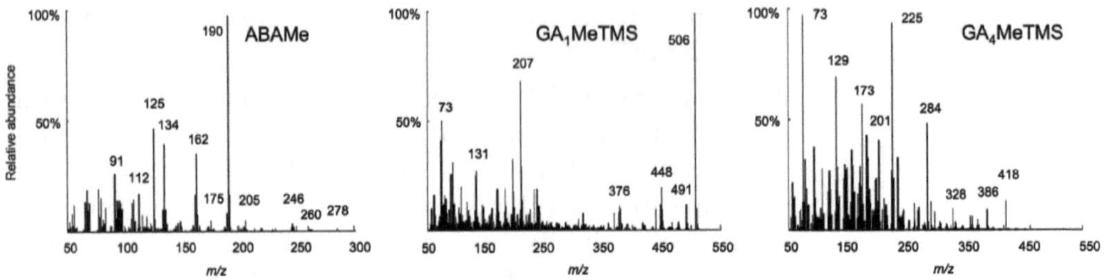

Fig. 5.2. Mass spectra of ABAMe and GA_1MeTMS and GA_4MeTMS. *X*-axis and *Y*-axis indicate *m/z* and relative abundance (%), respectively. Molecular weight of each fragment is shown with the corresponding peak.

settings are shown below. GA_1MeTMS, GA_4MeTMS, and ABAMe run off at 9 min 40 s, 8 min 25 s, and 7 min 00 s, respectively (*see* **Notes 13** and **14**). The mass spectra of ABAMe, GA_1MeTMS, and GA_4MeTMS are shown in **Fig.** 5.2.

2. GC-MS settings.
   a. Injection volume: 1–5 μL.
   b. Injection temperature: 250°C.
   c. Splitteless mode.
   d. GC oven gradient: The column temperature is maintained at 80°C for 1 min and increased to 245°C at rate of 30°C/min, and to 280°C at rate of 5°C/min.
   e. Interface temperature: 250°C.
   f. MS source temperature: 200°C.
   g. GC carrier gas flow: 1.0 ml/min, high-quality helium gas: 99.99995%
   h. Ionization energy: potential 70 eV, current 300 μA.
   i. MS analysis: Electron Impact (EI) ionization mode, selected ion monitoring (SIM).
3. The following mass-to-charge ratio peaks are used for quantification: for ABA, 190; [$^2H_6$]ABA, 194; GA_1, 506; [$^2H_2$]GA_1, 508; GA_4, 284; and [$^2H_2$]GA_4, 286. Backup ions (ABA, 260; [$^2H_6$]ABA, 266; GA_1, 448; [$^2H_2$]GA_1, 450; GA_4, 418 and [$^2H_2$]GA_4, 420) are also monitored for peak confirmation (*see* **Note 15**).

4. Notes

1. The extraction solution can be substituted with 80% (v/v) acetone. Ten milliliters extraction solution is added per 1 g FW of plant material.

2. [$^{2}H_6$]ABA can be purchased from ICON SERVICES (Summit, NJ, USA). [$^{13}C_2$]ABA can be obtained from Prof. T. Asami (The University of Tokyo) *(1)*. Deuterium-labeled GA_1 and GA_4 can be purchased from Prof. L.N. Mander of the Australian National University.
3. The volume of C18 cartridge column used is 1 cc_100 mg for ABA measurement, and 12 cc_2 g for GA measurement. For the DEA columns, column volumes are 1 cc_100 mg for ABA measurement and 3 cc_500 mg for GA measurement.
4. Diazomethane is toxic and explosive and should be handled with special care in a fume-hood. Avoid the use of sharp glass and physical agitation of the diazomethane container. Diethyl ether is highly flammable (boiling point: 34.5°C). Therefore, ethereal diazomethane solution needs to be stored in a tightly sealed glass bottle in an explosive-proof refrigerator. A yellowish colored solution is indicative of active diazomethane; this color disappears gradually during storage.
5. Typical amounts of plant material for measurements of ABA and GAs are shown in **Table 5.1**. The required amount of plant material is dependent on both purification efficiency and on the sensitivity of MS, but can be reduced with experience.
6. The amount of the internal standards to be added is estimated by semi-quantitative ELISA method or surveying the literature that describes hormone levels in similar plant materials. The range of quantitative detection of endogenous and internal standard compounds is estimated by making a calibration

Table 5.1
Examples of amounts of added standard compounds

Plant Material	Amounts of internal standards added/amounts of plant material (content)		
	ABA	GA_1	GA_4
Turgid Arabidopsis shoot	0.5–2 ng/0.5 gFW (1–4 ng/gFW)	20 ng/15 gFW (1.3 ng/gFW)	20 ng/15 gFW (1.3 ng/gFW)
Drought stressed Arabidopsis shoot	1–5 ng/0.1 gFW (10–50 ng/gFW)	20 ng/15 gFW (1.3 ng/gFW)	20 ng/15 gFW (1.3 ng/gFW)
Turgid tomato shoot	10–20 ng/0.1 gFW (100–200 ng/gFW)		
Drought stressed tomato shoots	100–200 ng/0.1 gFW (1000–2000 ng/gFW)		
Rice seedling	1–5 ng/1 gFW		

curve prior to quantification of endogenous hormones. **Table 5.1** shows examples of amounts of internal standards to be added. The amount of ABA and its catabolites in *Arabidopsis* tissues is described in *(2)*.

7. For purification of the ABA fractions, the aqueous phase can be cleaned up by several rounds of extraction with 5 ml of *n*-hexane. This procedure allows the omission of the cartridge column steps.
8. In the case of ABA extraction, ethyl acetate can be substituted with diethyl ether in solvent partitioning. Collected diethyl ether fractions are dried under an N_2 stream at 30°C and directly subjected to HPLC fractionation.
9. The HCl solution should be added gradually, because CO_2 gas is generated in this step.
10. The C18 cartridge column can be substituted with an Oasis HLB cartridge column, which allows the stable recovery of the target compounds.
11. Because MSTFA is sensitive to high humidity or water, samples have to be completely dried. After derivatization, seal the glass vial tightly and store it in a freezer.
12. The GA_4 fraction often requires another round of HPLC separation using an anion exchange column (Senshu Pak $N(CH_3)_2$, 4.6 mm i.d. × 150 mm) eluted with 0.1% (v/v) acetic acid in methanol at a flow rate of 1 ml/min. The column temperature is held constant at 40°C. Fractions between 11 and 13 min are collected and dried.
13. The retention time of isotope-labeled ABA and GAs is slightly shorter than non-labeled endogenous compounds (e.g., 0.2–0.5 s) due to the isotope effect.
14. Diazomethane methylates ABA at the carboxylic group, but not the hydroxyl group.
15. For accurate identification and quantification of chemicals, it is important to monitor multiple pairs of ions from the internal standard and the endogenous compound.

References

1. Asami, T., Sekimata, K., Wang, J. M., Yoneyama, K., Takeuchi, Y. and Yoshida, S. (1999) Preparation of (±)–[1,2–$^{13}C_2$] abscisic acid for use as a stable and pure internal standard. *J Chem Res* (Synop) 11, 658–659.
2. Priest, D. M., Ambrose, S. J., Vaistij, F. E., Elias, L., Higgins, G. S., Ross, A. R., Abrams, S. R. and Bowles, D. J. (2006) Use of the glucosyltransferase UGT71B6 to disturb abscisic acid homeostasis in *Arabidopsis thaliana*. *Plant J* 46, 492–502.

Chapter 6

Pull-Down Assays for Plant Hormone Research

Stefan Kepinski

Abstract

Hormonal signals are transduced (and sometimes perceived) by protein–protein interactions. Understanding these interactions is therefore crucial to understanding the network of signalling components as a whole. Often, genetic analysis serves up a selection of players that may or may not interact directly to carry the signal in question and further insight inevitably leads to some sort of biochemistry. Using the example of the auxin-regulated interaction between the auxin receptor TIR1 and the Aux/IAA repressor proteins, this chapter deals with some of this biochemistry, describing a very simple assay for looking at which proteins are interacting, and if and how those interactions are regulated.

Key words: Pull-down assay, Auxin, 2,4-D, SCF, TIR1, Aux/IAA.

1. Introduction

A simple pull-down method of analysing protein–protein interactions has been central to unravelling the early events of auxin perception and signalling *(1)*. The essence of the pull-down assay is that one binding partner is immobilised on to one of a variety of resins or beads which can then be recovered by centrifugation, allowing the extent of co-purification of the other partner to be assayed. It is an in vitro technique and one that has been extremely powerful in dissecting how the auxin signal is perceived and transduced. Genetic analysis had identified both a family of extremely unstable transcriptional repressors called Aux/IAAs and a number of components of the ubiquitin-mediated protein degradation system, notably the ubiquitin-ligase component TIR1, as being important in auxin signalling *(1, 2)*. Further molecular genetic experiments showed that Aux/IAAs were destabilised in response to auxin and linked Aux/IAA stability to TIR1 function *(1–3)*.

Sean Cutler, Dario Bonetta (eds.), *Plant Hormones: Methods and Protocols, Second Edition, vol. 495*

DOI 10.1007/978-1-59745-477-3_6 Springerprotocols.com

In order to test the conspicuous possibility that TIR1 might recruit Aux/IAAs for ubiquitination and hence degradation, a pull-down assay was developed in which Aux/IAAs were expressed as GST-fusion proteins which could then be used to 'pull-down' proteins from an extract of a transgenic *Arabidopsis* line expressing a myc-tagged derivative of TIR1 (**Fig. 6.1A**).

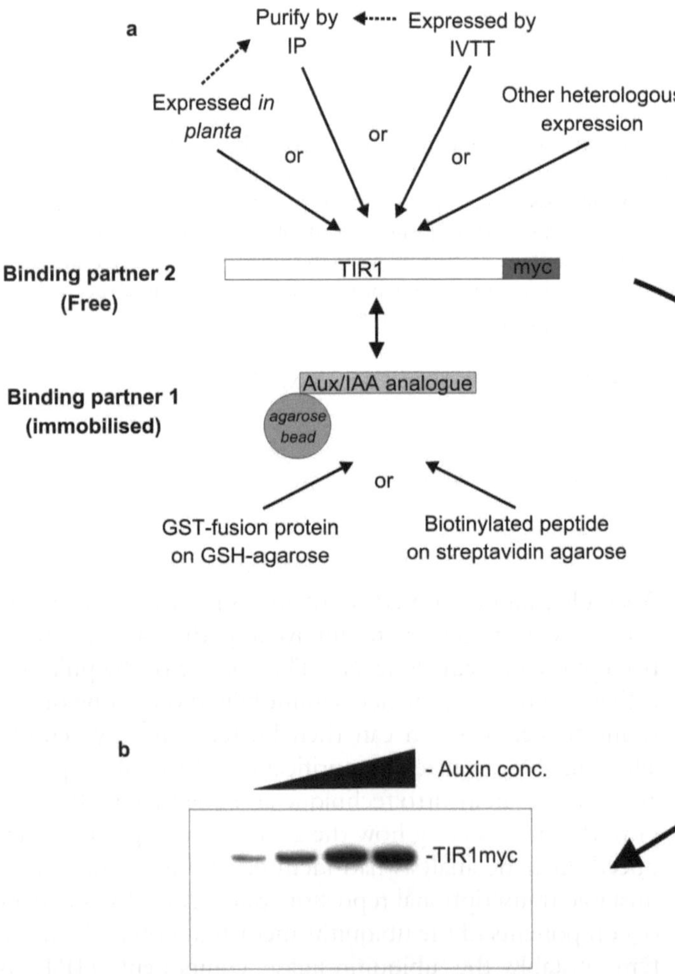

Fig. 6.1. (**A**) The basic format of the pull down assay. One binding partner is immobilised onto agarose beads, while the other is free. Here the immobilised partner is the Aux/IAA, in the form of either a GST-Aux/IAA fusion protein (e.g. GST-AXR2) on glutathione (GSH)-agarose beads or a synthetic biotinylated peptide (on streptavidin agarose beads) encompassing the 13 amino acids from AXR2 which are necessary and sufficient for auxin-regulated interaction with TIR1. The free partner is TIR1, as a myc-tagged derivative which can then be detected by anti-myc western blot (b). TIR1myc can be expressed and introduced into the pull-down reaction in several ways. Expression of functional TIR1 protein has been demonstrated in *Arabidopsis*, wheat germ in vitro translation reactions, and in other heterologous systems such as *Xenopus* embryos and insect cells. The pull-down reaction can either be performed directly with crude extracts or after partial purification of TIR1myc by anti-myc immunoprecipitation. (**B**) Anti-myc western blot showing a pull-down experiment testing the dose–response of Aux/IAA-TIR1 interaction (biotinylated domain II peptide against TIR1myc) to increasing concentrations of the auxin indole-3-acetic acid IAA (0, 0.5, 5, 50 μM). Each lane represents an individual pull-down reaction.

Using anti-myc western blots to monitor the recovery of TIR1myc, this method allowed the demonstration of an auxin-enhanced interaction between TIR1 and Aux/IAAs (**Fig. 6.1B**) that was consistent with the effect of auxin on Aux/IAA stability *(1)*. Further, the assay confirmed that mutations known to stabilise Aux/IAA proteins also decreased their ability to interact with TIR1 *(1)*. More detailed variations of this basic assay contributed to the mapping of the Aux/IAA domain required for interaction with TIR1 and provided the experimental framework for testing hypotheses on the regulation of the TIR1-Aux/IAA interaction by auxin, leading ultimately to the discovery that TIR1 itself is an auxin receptor *(5–8)*.

This work with auxin highlights several useful features and applications of the pull-down method in terms of confirming and dissecting known or suspected interactions indicated by genetics or other interaction methods such as yeast-2-hybrid. It is important to note that although the technique is an in vitro one and so carries all of the potential drawbacks that go along with doing any biological research ex vivo, the pull-down method has some important advantages over in vivo methods such as yeast-2-hybrid. The principal benefit is that pull-down methods can be used to detect and analyse interactions which cannot occur in yeast because either post-translational modifications or other factors required to support the interaction are absent. This is because the flexibility of the pull-down assay allows more control over the method of expression of the protein partners. This flexibility also means that when coupled to mass spectrometry, pull-down methods can be a powerful companion to yeast-2-hybrid library screening when the aim is to identify novel interactors with a given 'bait' protein.

Although specific to the analysis of the TIR1-Aux/IAA interaction in auxin signalling, the methods described here should provide a toolkit which can used to study other protein–protein interactions both in plant hormone research and beyond.

2. Materials

2.1. Preparation of GST Fusion Proteins

1. BL21-Gold (DE3) competent cells (Stratagene, La Jolla, CA) (*see* **Note 1**).
2. 100 mM Isopropyl-1-thio-β-D-galactopyranoside (IPTG) (Sigma) (1000X stock, dissolved in water and sterilised by filtration, store at –20°C).
3. Luria Broth (LB) containing 100 μg ml^{-1} ampicillin: Dissolve 10 g tryptone, 5 g yeast extract, 10 g NaCl in 900 ml water,

adjust pH 7.0 and make up to 1000 ml with water. Sterilise by autoclaving and allow to cool to > 50°C. Just before use, add the appropriate volume of ampicillin from a 1000X stock (100 mg ml^{-1}, dissolved in water and stored at –20°C).

4. Phosphate buffered saline (PBS): Prepare 10X stock with 1.37 M NaCl, 27 mM KCl, 100 mM Na_2HPO_4, 18 mM KH_2PO_4 (adjust to pH 7.4 with HCl, if necessary) and autoclave before storage at room temperature. Prepare working solution by dilution of one part with nine parts water.
5. BugBuster reagent (Novagen).
6. Benzonase (Novagen).
7. Glutathione (GSH) sepharose 4B (GE Healthcare).
8. Inhibitors: 100 mM Phenylmethanesulfonyl fluoride (PMSF – serine protease inhibitor) (Sigma) (100X stock dissolved in isopropanol), 1 M Dithiothreitol (DTT – strong reducing agent) (Sigma) (1000X stock dissolved in 10 mM sodium acetate pH5.2). All stored at –20°C.
9. Triton X-100 (Sigma).
10. Sterile glycerol (Sigma).
11. General plasticware – 1.5 ml microfuge tubes, sterile 15 and 50 ml screw-cap conical tubes, large centrifuge containers (250–500 ml).

2.2. SDS-PAGE and Coomassie Staining

1. NuPAGE 4 × LDS loading buffer, NuPAGE 10 × Reducing agent, NuPAGE Novex 4–12% Bis-Tris gel (10 well), NuPAGE 20X MES buffer, NuPAGE antioxidant (Invitrogen).
2. Novex X-cell Gel Electrophoresis rig (Invitrogen), power supply capable of delivering 200 W.
3. MultiMark prestained molecular markers (Invitrogen).
4. SimplyBlue Safestain (Invitrogen).

2.3. Immunoprecipitation of Myc-Tagged TIR1

1. Extraction buffer (EB): 150 mM NaCl, 100 mM Tris pH 7.5, 0.5% Nonidet-P40 (NP40).
2. Inhibitors: 100 mM PMSF as above. 1 mg ml^{-1} Pepstatin A (inhibitor of aspartic proteinases) (Sigma) (1000X stock dissolved in ethanol) 10 mM MG132 (peptide inhibitor of the 26S proteasome) (Calbiochem) (1000X stock dissolved in dimethyl sulfoxide). All stored at –20°C.
3. Liquid Nitrogen (^{L}N).
4. Pestle and mortar, Centrifuge, syringe-mounting 0.22 μm filtration disc (Acrodisc or similar), sterile 50 ml screw-cap conical tubes.
5. Anti-c-myc (9E10) affinity (Covance).

6. Sepharose 4B (GE Healthcare).
7. Empty 1 ml chromatography column (Isolute 1 ml Filtration column, part no. 120-1021-A, Biotage, Sweden or similar).
8. 100 mM glycine pH 2.9.

2.4. Pull-Down Assay Using GST-AXR3 Protein Against Crude Plant Extract

1. Plant extract as described **Sections** 2.3/3.3 but also including 10 μM DTT (from a 1 M stock as above).
2. GST-AXR2 coupled to GSH-sepharose (i.e. from protocol 3.1).
3. Indole-3-acetic acid (IAA, Sigma).
4. Rotating mixer at 4°C (i.e. either in a cold room or cooled cabinet).
5. EB as described in **Section 2.3.**
6. NuPAGE 4 × LDS loading buffer as above.

2.5. Pull-Down Assay Using Biotinylated Aux/IAA Peptide Against Immunopurified TIR1myc

1. Immunopurified TIR1myc from protocol 3.3.
2. Amino-terminally biotinylated synthetic peptide of the sequence (Biotinyl-)AKAQVVGWPPVRNYRKN.
3. Streptavidin-agarose (Upstate/Millipore).
4. 1 × PBS as above.
5. EB as described in **Section 2.3.**
6. NuPAGE 4 × LDS loading buffer as above.

2.6. Western Transfer and Detection

1. Novex X-cell Gel Electrophoresis rig and X-blot module with blotting pads (Invitrogen).
2. NuPAGE 20X transfer buffer, NuPAGE antioxidant as above.
3. Methanol.
4. Invitrolon PVDF membrane (Invitrogen) or similar PVDF membrane.
5. Flat nosed forceps.
6. 1 × PBS containing 0.1% Tween 20.
7. Dried non-fat milk (e.g. 'Marvel' or similar).
8. Anti-c-myc (9E10) monoclonal antibody (Covance, MMS-150R, raw ascites fluid).
9. Goat anti-mouse IgG (γ-chain specific) HRP-conjugate (Sigma, A3673).
10. ECL Plus reagents (GE Healthcare).
11. Acetate/OHP sheets or clingfilm/Saranwrap.
12. Ponceau S protein stain (Sigma).

3. Methods

Given that the basic requirements of a pull-down assay are that one partner is tethered to an affinity matrix or resin and the free partner is detectable in some manner, there are numerous ways in which a successful pull-down assay can be designed (**Fig. 6.1**). The methods described in this chapter are those that have proved most useful in the analysis of the interaction between Aux/IAAs and TIR1 (and related AFB proteins) *(1, 5–9)*, beginning with the preparation of GST-Aux/IAA fusion proteins and the immunopurification of TIR1 and going on to pull-down assays themselves. A few simple variations and considerations which may broaden the applicability of these methods to the analysis of other protein–protein interactions are mentioned here and throughout the protocols and notes. These relate to the different ways in which the proteins can be expressed, tagged and detected (**Fig. 6.1**). For example, to immobilise the first binding partner it is common to express that protein as a GST-fusion protein (as described in detail below) *(1)*, but it is also possible to use other tags that can either be readily coupled to a resin (e.g. 6xHIS tag purified on Nickel resin) or recognised by an antibody (e.g. HA, FLAG, myc, etc.) which can then be immobilised on protein A/G agarose. If purified protein is obtainable then it is also possible to couple this via an amide linkage to NHS-activated sepharose (GE Healthcare) or similar. Of course, if specific antibodies are available for the proteins in question then one of these may be used to immunoprecipitate (as described in **Section 3.3**), from crude plant extracts, one endogenous partner while the co-purification of the other is monitored. Such an approach is attractive because only the interaction between endogenous unmodified proteins are being assayed, but can be limited by the often low abundance of the signalling components under investigation and the risk that antibody binding hinders the interaction between partners.

To overcome the problem of low abundance, it is often necessary to over-express one or both partners. In the most common incarnation of the pull-down assay, as in the analysis of GST-AXR2 and TIR1myc interaction described below, the bait partner is (over-) expressed heterologously and used against crude extracts overexpressing the second partner (in this case of a transgenic *Arabidopsis* TIR1myc line) (**Fig. 6.1**). Although it is usual to use extracts of the relevant species in order to capture as far as possible a representation of typical binding activity, there can be occasions where 'abstracting' the interaction is useful. The case of the TIR1-Aux/IAA interaction is an example where, to test the idea that auxin receptor activity might require no more than

TIR1, auxin, and an Aux/IAA, the heterologous expression (in frog or insect cells) or synthesis of both protein components was invaluable in excluding, beyond reasonable doubt, the possibility that other proteins were involved (7, 8).

3.1. Preparation of GST Fusion Proteins

The following protocol has been found to be suitable for the medium-scale expression of GST-AXR2 using an expression construct made with the pGEX-2T vector from GE Healthcare. The optimal expression of other proteins should be determined empirically (*see* **Note 3**).

1. From a single colony of BL21 cells containing the GST-AXR2 plasmid, inoculate 2 × 50 ml of LB medium containing 100 μg ml^{-1} ampicillin and incubate at 37°C with vigorous shaking for 15 h (usually overnight).
2. Dilute the overnight culture 1:10 into fresh LB medium containing 100 μg ml^{-1} ampicillin and incubate at 37°C with vigorous shaking for a further hour (*see* **Note 2**).
3. Add IPTG to a final concentration of 100 μg ml^{-1} and incubate at 30°C with vigorous shaking for a further 3–4 h (*see* **Note 3**).
4. To harvest the cells, the cultures should be centrifuged at 7700 × g for 10 min at 4°C. This is best done using large (250–500 ml) centrifuge containers and the appropriate rotor/centrifuge combination (e.g. Sorvall or Beckmann). Record the weight of each empty container so that the weight of the pellet can be determined later.
5. Discard the supernatant, drain the pellet and then place on ice. Wash briefly with ice-cold PBS (using 1/10 of the spun culture volume, swirling gently 4–5 times) and resediment the cells by centrifuging at 7700 × g for 2 min at 4°C. Discard the supernatant, drain the pellet and then place on ice.
6. For each gram of pellet, add 5 ml of BugBuster reagent and 5 μl of Benzonase with DTT to 5 μM and PMSF to 1 μM. Resuspend by vigorous pipetting. Transfer to 15 ml Falcon tubes and incubate at room temperature (RT) for 20 min with gentle mixing. Remove insoluble material by centrifugation at 16000 × g for 20 min at 4°C. Transfer the supernatant to a fresh tube and remove 5 μL for later SDS-PAGE analysis (*see* **Note 4**).
7. To begin batch purification of the GST-fusion protein, add 400 μL of a 50% suspension of GSH-Sepharose to each sample tube (assuming an original *E. coli* culture volume of 500 ml). The GSH-Sepharose resin is prepared by washing the required volume of original stock slurry in 10 bed volumes (bed volume = volume of sedimented resin) of ice-cold PBS to remove the ethanol in the storage solution.

Gently centrifuge the resin (500 × g, 5 min) to sediment and then resuspend in PBS as 50% slurry.

8. Incubate sample with gentle mixing at 4°C for 30 min and then centrifuge (500 × g, 2 min) to sediment the resin. Remove the supernatant and wash the pellet in 10 bed volumes of PBS containing 0.5% Triton X-100, 1 μM PMSF and 10 μM DTT for 5 min at 4°C. Sediment by centrifugation as before and repeat for a total of three washes.
9. Resuspend the resin in an equal volume of PBS containing 10% glycerol and remove 5 μL for SDS-PAGE analysis. Divide the remainder into 2 aliquots and store at −80°C (*see* **Note 5**).
10. After SDS-PAGE analysis (*see* **Section 3.2**), as well as a strong band of the expected size, it is common to observe a number of smaller degradation products. Provided the GST-fusion protein band dominates, this is no cause for concern for the purposes here (*see* **Note 6**).

3.2. SDS-PAGE and Coomassie Staining

The following protocol assumes the use of the Invitrogen NuPage Novex gel electrophoresis system (Invitrogen) run under denaturing conditions, although there are several companies supplying suitable PAGE gel equipment.

1. Set a heating block to 70°C. Make up each sample for electrophoresis by adding the appropriate volumes of 4X LDS loading buffer and 10X reducing agent (up to a maximum volume of 25 μl). Incubate the samples at 70°C for 10 min.
2. While the samples are incubating, make up 800 ml of 1X MES running buffer with deionised water. Aliquot 200 ml of this and add 500 μL of antioxidant solution. Take a precast 10-well NuPAGE Novex 4–12% Bis-Tris gel, remove the comb and wash the wells with 2–3 ml of the 1X MES containing antioxidant. Assemble the gel within the Xcell Surelock electrophoresis rig (Invitrogen) and fill the upper chamber with 1X MES running buffer containing the antioxidant and the lower chamber with the rest of the 1X MES running buffer.
3. Load each sample (up to 25 μL) and prestained molecular weight markers (Invitrogen MultiMark or similar). Be sure to load marker lanes asymmetrically to aid later orientation of the gel. Connect the power supply and run the gel for 38 min at 200 V.
4. Disassemble the gel rig and remove the gel by carefully breaking open the gel cassette. Unless performing Western transfer (*see* **Section 3.6**), carefully slide the gel into a small container containing 100 ml or so of deionised water. These gels are relatively robust but still require gentle handling. Use a sharp pallet knife or similar to remove the wells (which can otherwise catch and cause tearing). Agitate gently at RT for 5 min, pour off the water and then repeat for a total of three water washes.

5. After pouring off as much of the final wash as possible, add sufficient SimplyBlue SafeStain (or similar colloidal coomassie stain) to cover the gel. Incubate for 1 h at RT with gentle agitation. Discard the stain and replace with 100 ml of deionised water (select a container that allows this volume to cover the gel). Incubate for a further hour to clear the background. To clear the background completely, wash once more for several hours in 100 ml deionised water.

3.3. Immunoprecipitation of Myc-Tagged TIR1 (or Other Proteins)

This protocol describes a method for purifying myc-tagged proteins for subsequent pull-down analysis from plant extracts using anti-c-myc antibody, although it can be used for immunopurification of tagged proteins generated by other methods (e.g. heterologous or in vitro expression, *see* **Section 3.4**). The method assumes the availability of a suitable source of myc-tagged protein, in this case a transgenic *Arabidopsis* line expressing a myc-tagged derivative of TIR1 (TIR1myc) but can easily be adapted for other proteins or proteins tagged with different epitopes (e.g. HA, HIS, FLAG) by substituting in the relevant antibody type.

1. Begin by preparing an extract of 600–800 7- to 10-day old TIR1myc (or other) seedlings. These numbers of TIR1-myc plants can be conveniently grown in liquid culture (*see* **Note 7**). Plants are harvested by draining off the liquid media (use a tea strainer or similar) before being well blotted between several sheets of laboratory tissues (press firmly to remove most of the media). Work quickly while harvesting and get the tissue into a suitably sized mortar containing liquid nitrogen (^{L}N) without delay. Grind hard with a pestle until a fine powder is formed (*see* **Note 8**) and transfer to 2-3 precooled 50 ml graduated Falcon/Greiner tubes (so that each tube is roughly half full with ground tissue).
2. Estimate the bed volume of powder in each tube and add 0.75 volumes of extraction buffer (EB: 150 mM NaCl, 100 mM Tris pH 7.5, 0.5% NP-40) containing 1 mM PMSF, 1 μg ml^{-1} pepstatin A, 10 μM MG132. Note that unlike other protocols in this chapter, this EB specifically excludes DTT so as to avoid disrupting the disulfide bridges holding the antibody heavy and light chains together. Use a spatula to mix and help wet the powder through as quickly as possible, forming a sorbet-like as mass. Close the tubes and shake vigorously to thaw the mixture. Once the lysate is running freely transfer to appropriate centrifuge tubes, balance and then centrifuge at 16,000 × g at 4°C for 15 min. Filter the supernatant through a 0.22 μM syringe filter. Extractions performed in this way will typically yield total protein concentrations in the range of 2.5–3.5 mg ml^{-1} (*see* **Note 9**). Set aside ~ 24 mg crude extract

(7–9 ml of extract) for the immunoprecipitation (IP) and snap-freeze the remainder in ^{L}N for storage at –80°C.

3. Preclear the extract by adding 800 μL (bed volume) of Sepharose 4B and incubating at 4°C for 1 h with mixing. Sediment the Sepharose (3000 × g, 4°C, 5 min) and transfer the supernatant to a new tube.
4. Wash 350 μL of anti-myc affinity matrix in 10 ml of ice-cold EB for 1 min with constant mixing. Sediment (500 × g, 5 min), discard the wash and then add the precleared extract and incubate at 4°C for 2 h with gentle mixing. Sediment the matrix (500 × g, 5 min, 4°C) and wash with 12 ml of ice-cold EB. Sediment as before and repeat the wash for a total of three washes. After the final wash, sediment the matrix and discard the majority of the wash, leaving a volume roughly equal to the matrix volume. Using an 1 ml micropipettor, gently resuspend the matrix in the remaining wash and quickly transfer the suspension to an empty 1 ml chromatography column. Place the column in a fresh 15 ml conical tube sitting in ice and gently pack the column by centrifuging the column and tube (use a centrifuge with a swing-out rotor for all column spins) at 500 × g, for 1 min (*see* **Note 10**). Add 0.75 ml of ice-cold EB to the tube from which the matrix suspension has been removed and use this volume to wash the bottom of the tube to collect any remaining matrix. Transfer to the column.
5. To elute the myc-tagged protein, spin the column briefly (500 × g, 1 min) and place in to a clean collection tube before immediately adding 400 μL of EB containing 400 μg ml^{-1} c-myc peptide. To perfuse the matrix with eluant, insert a 1 ml micropipette tip into the top of the column (forming a seal) and apply brief and gentle pressure via the micropipettor. Let the column stand for 10 min at 4°C. Recover the eluate by centrifugation at 500 × g for 2 min and reapply to the column, perfusing as before. After another 10 min at 4°C, recover the eluate by centrifugation at 500 × g for 5 min. Aliquot 5 μL for western analysis (*see* **Section 3.6**) and store either on ice (very short term) or snap-freeze in ^{L}N and place at –80°C (long term).
6. To regenerate the matrix, pass 25 ml of 100 mM glycine pH 2.9 through the column followed by 10 ml of PBS. Store the matrix in the column in PBS at 4°C until further use. The matrix can be recovered by inverting the column and using a pipette to wash through with PBS.

3.4. Pull-Down Assay Using GST-AXR3 Protein Against Crude Plant Extract

As mentioned above, the flexibility of the pull-down assay means that there are often several ways in which a protein–protein interaction can be tested. For clarity, two pull-down formats for assaying the auxin-enhanced interaction between Aux/IAAs and TIR1 will be described in this chapter. The first (3.4) uses a GST-AXR2 fusion protein and crude extracts of the transgenic *Arabidopsis* line expressing myc-tagged TIR1 described in **Section 3.3**, while the second (3.5) uses a short synthetic Aux/IAA peptide and TIR1-myc either partially purified by IP (3.3). It is important to note that both GST-AXR2 or biotinylated domain II peptide could be used successfully with either crude extract or immunopurified TIR1myc to allow a successful assay and as stated before, all components could be substituted for those relevant to whichever biological interaction is under investigation. If this interaction is uncharacterised, it is essential that more extensive controls are employed, using for example, unfused-GST protein, random or mutant peptides as appropriate.

The following protocol can also be readily modified to allow the analysis of auxin binding in the TIR1-Aux/IAA interaction using tritiated IAA (*see* **Note 11**).

1. Prepare extracts as in **Section 3.3** (**steps 1** and **2**), the only exception being that 10 μM DTT should be included in the EB. The pull-down assay described here typically requires 500–1000 μL of crude extract per assay so the numbers of plants suggested in **Section 3.3** will be sufficient for several assays. From a single batch of extract, dispense into 1.5 ml microcentrifuge tubes aliquots of ~2.5 mg crude extract for each assay required (*see* **Note 12**). Keep the reaction tubes on ice.

2. Thaw on ice the GST-AXR2 protein coupled to GSH-sepharose from 3.1, vortex very briefly to resuspend and dispense 10 μL into each reaction tube. Work quickly, using the pipette tip and pipetting up and down to keep the suspension evenly suspended. The idea is to ensure that equal amounts of GST-AXR2 resin are dispensed to each reaction (*see* **Note 13**). This can be difficult to judge before SDS-PAGE analysis but it is worth spinning the tubes briefly in a microfuge (4–5 s) and checking that the pellet sizes look similar. If they do not, recover the extracts avoiding carrying over the GST-AXR2, pool and mix them, and then re-aliquot.

3. Add the auxinic treatments as appropriate. For a simple IAA dose response using 0, 0.5, 5, 50 μM IAA for example, make dilutions (in 80% ethanol) from a 50 mM stock of IAA dissolved in 80% ethanol such that the treatment is administered to each reaction in 2 μl being sure to add 2 μL of 80% to the mock/control. Incubate the reactions at 4°C for 30 min with gentle end over end mixing (using a daisy wheel rotator or similar) (*see* **Note 14**).

4. Wash the GST-AXR2 resin by spinning briefly (10 s in a microfuge) to sediment, discarding the extract and replacing with 1 ml of ice-cold EB incubate at 4°C for 4 min with

mixing. Sediment again and repeat the 4 min wash twice more for a total of three washes. At each wash step, take care not to remove any beads. After the final wash has been removed, spin briefly again and remove as much wash as possible using a fine pipette tip (*see* **Note 15**).

5. Elute the protein from the beads by adding 20 μl of hot (heated to 70°C) 1 × LDS/reducing agent loading buffer (as in **Section 3.2**). Place the tubes at 70°C for 10 min and load the sample (including beads) onto a PAGE gel as described in **Section 3.2**. Once the gel has run proceed immediately to western transfer and detection (3.6).

3.5. Pull-Down Assay Using Biotinylated Aux/IAA Peptide Against Immunopurified TIR1myc

If the interaction domain of at least one of the binding partners is defined and reasonably small (< ~50 amino acids), that partner can be synthesised as a biotinylated peptide. Provided the peptide exhibits binding typical of the full length protein this has a number advantages for pull-down assays including consistency of a highly purified molecule and the fact that the once coupled to streptavidin-agarose via the biotin tag, the peptide remains bound under all but the harshest elution conditions. The peptide used here encompasses the core amino acids from domain II of the Aux/IAA protein AXR2/IAA7, which are necessary and sufficient for auxin-regulated interaction with TIR1. The protocol describes a pull-down where the second partner (TIR1myc) has already been partially purified by IP (3.3). It is also possible to synthesise either one or both partners by in vitro transcription/translation (IVTT) in wheatgerm lysate systems (*see* **Note 16**). If the functionality of the proteins can be established, this can be a very quick and flexible way of expressing proteins for in vitro interaction assays.

1. Equilibrate 1 ml (bed volume) of streptavidin agarose by washing in 10 ml of ice-cold 1 × PBS for 5 min (use a 15 ml tube). Centrifuge at 500 × g for 2 min to sediment and discard the wash. Add another 1.5 ml of 1 × PBS. Resuspend the peptide in sterile water at 1 mg ml^{-1}, vortexing briefly until fully dissolved. Add 400 μg of peptide (400 μl) to the washed streptavidin agarose (store unused peptide solution at –20°C) and incubate at RT for 30 min with constant mixing (*see* **Note 17**). Centrifuge as before, discard the peptide solution and add 14 ml of 1 × PBS. Incubate at RT for 5 min with mixing and sediment by centrifugation as before. Repeat this wash cycle for a total of three washes. Resuspend the resin in an equal volume of 1 × PBS (i.e. to make 2 ml of a 50% suspension). Store on ice or at 4°C.
2. Assemble the reactions by adding 425 μl of EB containing 1 mg ml^{-1}, 1 mM PMSF, 1 μg ml^{-1} pepstatin A, 10 μM MG132, and 10 μM DTT to a 1.5 ml microfuge tube. Add 75 μl of the immunopurified TIR1myc from 3.3 and 60 μl of

the 50% peptide-streptavidin agarose suspension prepared earlier (use the same technique as in point 2 of **Section 3.4** to ensure even distribution of peptide).

3. Add mock and hormonal treatments as appropriate and incubate the reactions at 4°C for 30 min with gentle end over end mixing.
4. Wash the peptide-streptavidin resin by spinning briefly (10 s in a microfuge) to sediment, discarding the extract and replacing with 1 ml of ice-cold EB incubate at 4°C for 4 min with mixing. Sediment again and repeat the 4-min wash twice more for a total of three washes. Because the volume of beads is greater than in 3.4, the pulled-down proteins should be eluted by placing the peptide-streptavidin resin in a small chromatography column (*see* **Note 15**).
5. Elute the protein from the beads by adding 20 μl of hot (heated to 70°C) 1 × LDS/reducing agent loading buffer (as in 3.2). Place the tubes at 70°C for 10 min and load the sample onto a PAGE gel as described in 3.2. Once the gel has run proceed immediately to western transfer and detection (3.6).

3.6. Western Transfer and Detection

The following protocol assumes that a gel has been run as in 3.2 and that the X-blot module of the NuPage Novex gel electrophoresis system (Invitrogen) is available. This system allows the efficient transfer of proteins to membrane in an hour at RT.

1. While the gel is running, prepare for the transfer by making up 1 l of 1 × NuPAGE transfer buffer containing 100 ml of methanol and 1 ml of NuPAGE antioxidant. Soak four blotting pads in this transfer buffer.
2. Remove one 'sandwich' of Invitrolon PVDF between the supplied filter paper sheets. Soak the filter paper sheets in transfer buffer and leave immersed. Wet the PVDF membrane in a small volume of methanol for 30 s and then immerse in transfer buffer to soak (the membrane should be transferred with clean flat-nosed forceps and will need to be agitated to get it to remain immersed in transfer buffer).
3. Once the gels have run, remove from the rig and carefully break open the gel cassette. Ensure that the gel remains on the larger plate. Place a presoaked piece of filter paper onto the gel (trying to avoid air bubbles), turn the cassette over and use the supplied gel knife to push the foot of the gel downwards, allowing the gel and filter paper to be released from the larger plate onto a clean area of bench. Immediately place the prewet PVDF membrane onto the gel (again, trying to avoid air bubbles), followed quickly by the second presoaked piece of filter paper. To ensure that there are no air bubbles between

the gel and membrane (which would prevent protein transfer), use a glass rod or pipette to apply gentle pressure with a 'rolling pin' action. Place two soaked blotting pads into the cathode tray of the blot module (the larger of the two components) and carefully place the filter paper/gel/PVDF sandwich onto these pads, ensuring that the gel is toward and the membrane away from the base of the cathode unit. Place the remaining soaked pads on top of the gel/membrane sandwich and then the anode core on top of that and assemble into the Xcell rig. Slowly pour 1 × transfer buffer over the pad/gel/membrane assembly until they are covered to a depth of around 1 cm. Fill the outer chamber with deionised water, connect the power supply and run the gel for 60 min at 30 V.

4. Turn off the power supply and remove the blot module. Remove the top two blotting pads and carefully lift the corner of the membrane to check that the prestained markers have transferred to the membrane (if they have not, reassemble the rig and re-run checking that the setting are correct and that there is a current). Remove the PVDF membrane and keeping the protein side uppermost, place in a small dish (just slightly larger than the size of the blot) and cover with 1 × PBST (PBS containing 0.1% Tween 20). If analysing a GST-Aux/IAA fusion-based pull-down, stain the blot with Ponceau S to check GST-Aux/IAA levels (*see* **Note 13**).
5. Block the membrane by incubating the membrane in 5% dried non-fat milk (e.g. Marvel brand or similar) in 1 × PBST for at least an hour with gentle agitation. Alternatively, the membrane can be blocked overnight at 4°C.
6. Wash the membrane twice briefly (50 ml of 1 × PBST, 2 min RT). Pour off the wash and replace with 10 ml of PBST. Add the primary anti-c-myc antibody at the appropriate dilution. For the antibody from Covance listed in **Section 2.6** this is usually in the region of 1:1000 but it is important to stress that this should be determined empirically (*see* **Note 18**). Incubate at RT for 90 min with gentle agitation. Pour off the antibody solution and rinse the blot with a few ml of 1 × PBST, pour that off and replace with 50 ml of 1 × PBST. Wash for 15 min at RT with gentle agitation then discard the wash and replace with another 50 ml of 1 × PBST and wash for a further 5 min.
7. Discard the wash and replace with 10 ml of 1 × PBST. Add the secondary HRP-conjugate antibody at the appropriate dilution. For the goat anti-mouse HRP-conjugate listed in **Section 2.6**, this is typically in the region of 1:5000. Incubate for 1 h at RT with gentle agitation. Pour off the antibody solution and rinse the blot with a few ml of 1 × PBST, pour that off and replace with 50 ml of 1 × PBST. Wash for 15 min at RT with gentle agitation then discard the wash and

replace with another 50 ml of 1 × PBST and wash for a further 5 min. Repeat this 5 min wash three more times.

8. While the final washes are being performed, aliquot 1 ml of ECL Plus Reagent A (from the larger bottle) into a 1.5 ml microfuge tube, cover with a small piece of foil and allow to warm to RT. Also allow Reagent B to warm to RT. To begin the detection reaction, add 25 μl of ECL Reagent B to the aliquot of Reagent A and prepare a pipettor with tip to pipette 1 ml of the mixed reagent. Lift the blot out of the final wash using flat-nosed forceps and holding it vertically, lay one edge against a paper towel for a few seconds to drain excess wash. Lay the blot on a clean piece of acetate film (or saranwrap, etc.) and immediately pipette a mixture of ECL reagents A and B onto the blot, ensuring that the whole membrane is covered. Incubate for 5 min and then drain off the ECL Plus using the paper towels as before. Place the blot on a new piece of acetate film. Immediately place another piece of acetate on top of the blot, starting at one end and laying it down gradually to avoid trapping air bubbles.
9. Move to a darkroom equipped with a safelight. Place the blot in an autoradiograph cassette and under safelight conditions expose to a piece of X-ray film (e.g. X ECL, GE Healthcare) for 20 s. Develop the film and assess the exposure. Perform further exposures so that the bands can be seen against a clear background. A typical result is shown in **Fig. 6.1B** showing the effect of auxin on the recovery of TIR1myc by pull-down with biotinylated domain II peptide.

4. Notes

1. A number of *E. coli* strains are suitable for protein expression. Ease of induction and high expression levels make the BL21(DE3) strain a good choice.
2. For fusion proteins other than GST-AXR2, use a spectrophotometer to check that the A_{600} is between 0.5 and 2 OD units.
3. Growing the cultures at 30°C rather than 37°C increases the yield of soluble protein with this particular fusion protein. For other proteins, it may well be necessary to experiment with induction conditions by varying temperature, length of induction, density of the culture at induction start, and IPTG concentration in order to maximise soluble protein (and minimise the formation of inclusion bodies).

4. BugBuster reagent and similar chemical lysis reagents offer a convenient method and breaking open the *E. coli* cells without need for expensive equipment. However, if a sonicator is available, users may wish to explore sonicating pellets resuspended in ice-cold PBS (5 ml per 100 ml of original culture volume) containing 0.5% Triton X-100 and 5 mM DTT.
5. For these pull-down protocols, it is useful to have the GST-fusion protein on the GSH-Sepharose beads. If free protein is required, the GST-fusion can be eluted by incubating the resin in an equal volume of 10 mM reduced glutathione in 50 mM Tris pH 8) for 10 min at RT. Centrifuge (500 × g, 5 min) and retain the supernatant (containing the eluted protein) for further use.
6. If the identity of any of the bands observed is in doubt, western blotting using anti-GST antibody (GE Healthcare) will help clarify the situation.
7. Growing large numbers of seedlings is most conveniently done in liquid culture. Begin by sterilising seeds by washing them for 30 min in 10% bleach, 0.01% Triton X 100, washing for a few seconds in 70% ethanol and rinsing thoroughly (5 or 6 times) in sterile distilled water before storing at 4°C for 2 days. Sow into 250 ml flasks (bunged with cotton wool) or bottles containing 60 ml of growth medium such as *Arabidopsis thaliana* salts *(10)* or Murashige and Skoog medium *(11)* at a rate of 100–200 seeds per bottle. Grow on a shaking platform under a 22°C/ 18°C 16 h/8 h light dark regime (80–100 μmol m^{-2} s^{-1}) for 7–10 days.
8. Good protein extractions depend on good tissue homogenization. To achieve this using a pestle and mortar and ^{L}N, grind hard in short bursts adding small amounts of ^{L}N frequently to keep the tissue as slurry. Keep grinding until a fine whiteish powder is formed and then tilt the pestle and add a final drop of ^{L}N which will collect the powder down into a 'cake' which can easily be scooped into a precooled tube using a spatula.
9. Measuring protein concentration. There are several methods. If a spectrophotometer capable of measuring abs at 595 nm is available (or at least between 570 and 610 nm) the Coomassie Plus system from Pierce is recommended because it tolerates the constituents of the buffer quite well. Nevertheless, the crude extract should be diluted 10 × in water before measuring the protein concentration against a set of standards between 0 and 800 μg ml^{-1}. Determine sample protein content by reference to the calibration curve derived from the assay of the standards.
10. Packing small columns. The idea is to pass excess wash through the column without drying the matrix out. Rather

than using centrifugation, it is also possible to pack the column adequately by inserting a 1 ml micropipette tip into the top of the column and applying brief and gentle pressure via the micropipettor. This has the advantage of allowing the passage of the mobile phase through the column to be observed so as to make sure the column remains wet.

11. It is possible to modify this GST-AXR2 based pull-down so that auxin binding rather than TIR1myc recovery is analysed. Assemble the reactions as stated but include a low concentration (0.01–0.1 μM) of tritiated (^{3}H) IAA (3-[5(n)-^{3}H] (Inodolylacetic acid, GE Healthcare) and as well as several increasing concentrations of unlabelled IAA or other auxins so that the displacement of the tritiated compound can be assessed. Wash the samples more briefly than for western analysis, using 2 × 1 min washes with 750 μl EB. Transfer the beads to scintillation vials containing appropriate scintillant (e.g. Ultima Gold Scintillation Fluid, Perkin Elmer) and perform scintillation counting using a scintillation counter. For this type of assay, GST-AXR2 on glutathione sepharose beads is better than Domain II peptide on streptavidin agarose beads because the former uses a lower volume of beads, affording a lower background (which arises as ^{3}H IAA is adsorbed onto the beads).

12. The experiment described here is based on several aliquots of a single extract type and seeks to establish the effect of adding auxin to otherwise identical reactions (same extract and peptide concentrations). Thus it does not matter if the total amount of crude extract in each reaction is slightly higher or lower than 2.5 mg ml^{-1} because the experiment is self-contained. However, if extracts are derived from different seedlings samples, for example where plants have been treated in some way prior to extraction, it is essential that the same amount and concentration of crude extract is used (which will require the dilution of more concentrated extracts using EB).

13. The even pipetting of sepharose resins is aided by using a pipette tip with 2–3 mm of the tip removed to increase its bore. When using GST-Aux/IAA fusions for pull-down assays, it is important to check that equivalent amounts of protein as used in each pull-down reaction. This is best done by staining the PVDF membrane to which the proteins have been transferred (3.6) using Ponceau S stain (Sigma). Just prior to western blotting, stain the membrane in a few ml of Ponceau S solution until the GST-AXR2 bands become visible. GST-AXR2 will be the most abundant protein in each lane. Rinse briefly with water to obtain a clearer background if required. Check that each lane has similar amounts of

GST-AXR2 protein and note any significant variation which should then be taken into account when assessing the recovery of TIR1myc after the western blotting is completed.

14. The 30-min incubation time is more than sufficient for the Aux/IAA-TIR1 interaction. If an uncharacterised interaction is being assayed, it would be sensible to extend this incubation time to 4 h and also run a GST only control to ensure that the interaction is specific to the GST fusion protein.

15. Removing the final wash without accidentally discarding beads can be difficult. To circumvent this, the final wash can be transferred to a small (~1 ml) chromatography column (similar to that used in **Section 3.3**). Use a pipettor to apply gentle pressure to clear the bulk of the wash and then spin briefly to remove the rest. Depending on the model of microfuge, this column can be placed in a 1.5 ml tube and spun in a microfuge (alternatively the column can be cut down to fit). To elute the protein, add 25 μl of hot (heated to 70°C) LDS loading buffer, incubate on the bench for 1 min and then spin the column in a fresh collection tube for 1 min.

16. In vitro transcription/translation in wheat germ lysates (or sometimes reticulocyte lysates – see comment below) can be an extremely quick and flexible way to express proteins for in vitro interaction analysis via pull-down assays. Proprietary kits are available (e.g. Promega and Roche) in which the transcription and translation reactions are coupled in one master mix meaning that the user need to only add either circular or linearised vector containing the gene of interest, with in-frame tags downstream of an SP6 or T7 promoter, depending on the particular kit in question. It is also possible to include labelled amino acids in order to label one of the binding partners which can be particularly useful if other methods of detection are not available for that protein (e.g. if no antibody is available or the protein does not tolerate epitope tagging). It is important to establish that protein produced in this way retains functionality. TIR1 is a good example of a protein for which caution is required in this regard. Expressed by IVTT in a wheat germ system (TNT® SP6 High-Yield Protein Expression System, Promega), TIR1 is functional while in a rabbit reticulocyte lysate system (TNT® SP6 Quick Coupled Transcription/Translation System, Promega,) it is not. One of the advantages of IVTT is that it is very easy to generate mutant protein derivatives by site directed mutagenesis (e.g. Quikchange, Stratagene) to establish which domains and residues are necessary/sufficient for the observed binding activity. To perform a pull-down using TIR1myc expressed in the TNT® SP6 High-Yield Protein Expression System rather than IP'd TIR1myc, substitute the 75 μl of the IP for

40 µl of IVTT reaction performed according to the manufacturer's instructions.

17. This is an excess of peptide to try to ensure that maximal occupation of biotin binding sites on the streptavidin. The excess is removed in the washes. It is possible to perform the assay by first adding the free biotinylated peptide to the reactions and subsequently capturing the peptide by adding sufficient streptavidin agarose later. Since the free peptide is easier to pipette precisely than peptide-streptavidin resin, this method may be preferred where small differences in pull-down binding are being assayed. However, this can increase the co-purification of endogenous biotin-containing proteins from the extracts.
18. Optimising antibody concentrations for western blotting. The correct dilution of antibodies depends on many factors including the specific activity of the antibody and it specificity for the antigen, the concentration of antigen (i.e. how much protein is there on the blot) and the method of detection (e.g. chemiluminescence, colourimetric, etc.). Perform optimisation experiments by producing several identical blots to test a range of dilutions for both primary and secondary antibodies.

Acknowledgments

The application of these techniques to TIR1-Aux/IAA interaction analysis was first developed in the Estelle Lab and I would like to thank Mark Estelle (Indiana University), Bill Gray (University of Minnesota) and Ottoline Leyser (University of York) for their advice and longstanding support.

References

1. Gray, W. M., Kepinski, S., Rouse, D., Leyser, O., Estelle, M. (2001) Auxin regulates SCF^{TIR1}-dependent degradation of AUX/IAA proteins. *Nature* 414, 271–276.
2. Woodward, A. W., Bartel, B. (2005). Auxin: regulation, action, and interaction. *Ann Bot* 95, 707–735.
3. Zenser, N., Ellsmore, A., Leasure, C., Callis, J. (2001) Auxin modulates the degradation rate of Aux/IAA proteins. *Proc Natl Acad Sci* USA 98, 11795–11800.
4. Tiwari, S. B., Wang, X.-J., Hagen, G., Guilfoyle, T. J. (2001) AUX/IAA proteins are active repressors, and their stability and activity are modulated by auxin. *Plant Cell* 13, 2809–2822.
5. Kepinski, S., Leyser, O. (2004) Auxin-induced SCF^{TIR1}-Aux/IAA interaction involves stable modification of the SCF^{TIR1} complex. *Proc Natl Acad Sci USA* 101, 12381–12386.
6. Dharmasiri, N., Dharmasiri, S., Jones, A. M., Estelle, M. (2003) Auxin action in a cell-free system. *Curr Biol* 13, 1418–1422.
7. Dharmasiri, N., Dharmasiri, S., Estelle, M. (2005) The F-box protein TIR1 is an auxin receptor. *Nature* 435, 441–445.

8. Kepinski, S., Leyser, O. (2005) The *Arabidopsis* F-box protein TIR1 is an auxin receptor. *Nature* 435, 446–451.
9. Dharmasiri, N., Dharmasiri, S., Weijers, D., Lechner, E., Yamada, M., Hobbie, L., Ehrismann, J. S., Jürgens, G., Estelle, M. (2005) Plant development is regulated by a family of auxin receptor F box proteins. *Dev Cell* 9, 109–119.
10. Wilson, A. K., Pickett, F. B., Turner, J. C., Estelle, M. (1990) A dominant mutation in *Arabidopsis* confers resistance to auxin, ethylene and abscisic acid. *Mol Gen Genet* 222, 377–383.
11. Murashige, T., Skoog, F. (1962) A revised medium for rapid growth and bioassays with tobacco tissue cultures. *Physiol Plant* 15, 473–497.

Chapter 7

Binding Assays for Brassinosteroid Receptors

Ana Caño-Delgado and Zhi-Yong Wang

Abstract

BRI1, a leucine-rich repeat (CRR)-receptor kinase protein, belongs to the plant hormone brassinosteroid (BR) receptor complex in *Arabidopsis thaliana* (Arabidopsis). The experiments described in this chapter to carry out BR-binding assays in Arabidopsis plants are largely based on previously published procedures by Wang et al. *(1)* and Caño-Delgado et al. *(2)*, which demonstrated that BRI1 and BRI1-like proteins BRL1 and BRL3, are components of the BR-receptor complexes *in vivo*. Furthermore, binding assays using a photo-affinity labelled castasterone have demonstrated that BRs are perceived directly by a subdomain of the extracellular domain of BRI1 defined by the 94 amino acids of the ID-LRR22 *(3)*.

Key words: BRI1, LRR-receptor kinase, ligand binding, Brassinolide, plant membrane protein microsomes.

1. Introduction

Brassinosteroids (BRs) are growth-promoting steroid hormones essential for normal plant growth and development. BR-deficient or insensitive mutant plants show extreme dwarfism and other defects in various developmental processes *(4)*. Molecular genetic studies have identified a number of genes required for BR responses in Arabidopsis *(5)*. These include the *BRI1* gene encoding a receptor-like kinase localized on the plasma membrane *(6, 7)*. Ligand binding assays performed with transgenic or mutant plants and with immunoprecipitated proteins demonstrated that BRI1 and its homologs are involved in BR perception *(1, 2)*. The photoaffinity labeling technique is a powerful chemical methodology for identifying the binding proteins of a specific small molecule and for the structural analysis of ligand binding domains. We have used BPCS, an analog of castasterone, a biosynthetic precursor of BL, which carries a

Sean Cutler, Dario Bonetta (eds.), *Plant Hormones: Methods and Protocols, Second Edition, vol. 495*

DOI 10.1007/978-1-59745-477-3_7 Springerprotocols.com

carbene-generating phenyldiazirine moiety and a biotin-tag. The phenyldiazirine moiety is expected to allow the formation of a covalent bond between BPCS and the binding region of specific proteins upon UV-irradiation, while the biotin-tag enables non-radioactive detection using an anti-biotin antibody. The use of photoaffinity crosslinking experiments using the BPCS, as herein described, led to the discovery that BRs are perceived directly by a subdomain of the extracellular domain of BRI1 defined by the 94 amino acids of the ID-LRR22 *(3)*.

2. Materials

2.1. Specialized Equipment for Radioassays

1. A radioactivity-controlled working area is needed to safetly handle [^{3}H]-labelled BL. This area should contain a benchtop microcentrifuge, a set of micropipettes, and a pair of forceps.
2. Glass microanalysis filter holders with a 25 mm friten glass support (Millipore®). Glass-fibre filters (Whatman, GF/F).

2.2. Solutions and Reagents for [^{3}H]-labelled BL Binding Assays

1. [^{3}H]-labelled BL, purchased from American RadioChemicals and prepared by tritium reduction of 25,26-dehydrobrassinolide, as shown in *(8)*.
2. BRI1-extraction buffer composition:

   50 mM Tris–HCl pH 7.5,
   10 mM NaCl,
   5 mM EDTA,
   1% Triton X-100,
   Protease Inhibitor cocktail (1/100 dilution).
3. BL-binding buffer composition:

   10 mM MES–KOH pH 5.7,
   5 mM $MgCl_2$,
   0.25 mM $CaCl_2$,
   0.25 mM Mannitol.
4. Membrane extraction buffer (MEB):

   20 mM Tris–HCl, pH 7.5,
   250 mM manitol,
   5 mM $MgCl_2$,
   0.1 mM CaCl,
   Protease inhibitor cocktail (see below).
5. Protease inhibitor cocktail (PIC, Sigma).
6. Polyclonal anti-GFP antibodies (Molecular Probes®).
7. 50% slurry immobilized protein A beads (Pierce®).

8. Readycast 4–20% polyacrylamide gels (Novex, Invitrogen).
9. SuperSignal Pico West chemiluminescent Substrate (Pierce).

2.3. Materials for Photoaffinity Crosslinking

1. BPCS compound [biotin-tagged photoaffinity castasterone, as described in *(3)*].
2. Biotin monoclonal antibody (Jackson ImmunoResearch).
3. Rabbit anti-mouse IgG conjugated to horseradish peroxidase (Bio-Rad).
4. T-TBS buffer: TBS buffer + 0.05% [w/v] Tween 20.
5. Blocking buffer: 0.05% [w/v] Tween 20, 5% [w/v] non-fat dry milk, 20 mM Tris–HCl pH 7.4, and 140 mM NaCl.
6. 24-Well plates (2 cm diameter).

3. Methods

3.1. [^{3}H]-BL-binding Assays in Inmunoprecipitated BRI1-GFP Fractions

3.1.1. Preparation of Protein Extracts

1. Harvest rosette leaves of 4 to 6-weeks-old BRI1-GFP transgenic Arabidopsis plants grown in 9 h light/15 h dark cycles. Wash the tissues carefully with distilled water and blot dry. Use extracts from Col-0 plants as negative control for the binding assays.
2. Weight the leaves and use 3–4 g per binding experiment.
3. Place the leaves in 500 ml centrifuge bottles. Keep them on ice and transfer to the 4°C working area.
4. Add 2 ml g^{-1} of chilled BRI1-extraction buffer and the protease inhibitor cocktail (8 ml BRI1 extraction buffer when using the suggested 4 g of tissue). Grind the tissue using a Polytron blender (medium speed) until completely homogenized (3–4 times for 1 min). Save a 20 µl aliquot for Western blot (WB) control.
5. Centrifuge the homogenized tissues at 18,000 rpm for 30 min at 4°C.
6. Filter the supernatant (~15 ml/sample) through two layers of Miracloth into 15 ml Falcon® Tubes (a 20 µl aliquot can be saved for WB control, A2).

3.1.2. Inmunoprecipitation (IP) of BRI1-GFP

1. To IP BRI1-like GFP-tagged proteins, add 1/3000 anti-GFP antibodies and 20 µl of 50% slurry immobilized-protein A beads per ml of extract (for a 15 ml IP volume use 5 µl anti-GFP antibodies and 300 µl of immobilized-protein A). The reaction can be carried in 15 ml volume tubes.
2. Incubate the IP by rocking at 4°C for 1 h.

3. Centrifuge the sample at 2000 rpm for 5 min. Take off the supernatant and transfer the beads to a clean 1.5 ml Eppendorf tube (save a 20 μl of supernatant for WB control, A3).
4. Wash the beads by additional centrifugation three times with 1 ml ice-cold BRI1 extraction buffer (no detergent added) 5 min at 3000 rpm.
5. Re-suspend the beads in 300 μl of BRI1-binding buffer and aliquot 100 μl/tube in Eppendorf tubes (save 30 μl beads for WB control, I1).

3.1.3. [^{3}H]-labelled BL Binding reaction

1. Incubate the IP fractions with different concentrations (2, 10, 25, 100, and 250 nM) of [^{3}H]-labelled BL, 1 mg ml^{-1} BSA for 30 min at RT. To determine non-specific background binding, use additionally 100-fold unlabelled BL (0.2, 1, 2.5, 10, 25 μM).
2. While incubation, mix the beads gently every 5 min.
3. Fill the scintillation tubes with 4 ml scintillation liquid, label them and place them in the counter racks (two scintillation tubes are needed per sample to separately count unbound and bound [^{3}H]-labelled BL).
4. Spin the binding reaction for 5 s. Add 10 μl supernatant in the scintillation tubes. This flow through serves to calculate the unbound [^{3}H]-labelled BL.
5. Wash the beads with BL binding buffer three times by centrifugation 1 min at 3,000 rpm (use the bench-top centrifuge placed in radioactive laboratory).
6. Re-suspend the beads containing the binding reactions in 4 ml scintillation counter liquid. Carefully use the micropipette to take all the beads from the tube, using scintillation liquid from the same sample can help.
7. Place the samples in the scintillation counter. Bound BL was quantified by scintillation counting.
8. Specific BL binding is calculated by subtracting the binding in the presence of 100-fold unlabelled BL from total binding measured without unlabelled BL. Import the raw binding data into the Kaleida graph® software (Synergy) that allows to calculate binding affinity, dissociation constant (K_d), and 50% Inhibitory Concentration (IC_{50}), and to produce graphic plotting of the results.
9. On the next day, separate the protein control samples A1, A2, A3 and I1 by SDS-PAGE loading the samples into 4–20% gels.
10. Western blot using anti-GFP antibody (1:5,000) to detect BRI1-GFP as described by *(9)*.

3.2. Microsomal Binding Assay Using Glass-Fibre Filters

3.2.1. Microsomal Preparation of Fresh Grinded Tissue

1. Harvest rosette leaves of 4 to 6-weeks-old plants grown in 9 h light/ 15 h dark. Wash the tissues with distilled water and carefully paper dry them.
2. Weight the leaves and use 3–4 g per binding experiment. Place them in 500 ml centrifuge bottles in an ice bucket.
3. Grind the tissue in 3 ml g^{-1} of chilled membrane-extraction buffer (MEB) and the protease inhibitor cocktail using a Polytron blender (medium speed) until completely homogenized (3–4 times for 1 min) (save a 20 μl aliquot for WB control, M1).
4. Filter the homogenized tissue through two layers of Miracloth (A 20 μl aliquot can be saved for WB control, M2).
5. Centrifuge the homogenized tissues at 10,000 × g for 10 min at 4°C. (A 20 μl aliquot can be saved for WB control, M3).
6. Ultracentrifuge the supernatant at 29,000 rpm (100,000 × g) for 1 h at 4°C.
7. Gently resuspend the microsomal pellet in BL-binding buffer. Quantify the microsomal concentration by Bradford assay according to manufacturer's instructions, and adjust protein concentration to 2mg ml^{-1} (a 20 μl aliquot can be saved for WB control, M4). At this step, samples can be frozen in liquid N_2 and stored at −80°C until needed.
8. Pre-soaking the glass-fibre filter membranes (Whatman, GF/F) in BL binding buffer containing 1 mg ml^{-1} BSA for at least 1 h at 4°C.
9. Aliquot 50 μl of microsomes (100 μg) in Eppendorf tubes and keep them at 4°C.

3.2.2. [^{3}H]-labelled BL Binding Reaction

1. In the radioactive laboratory, incubate 50 μl microsomes with different concentrations (0, 10, 25, 100 and 250 nM) of [^{3}H]-labelled BL, 1 mg ml^{-1} BSA. For background binding assays, use additionally 100-fold unlabelled BL. Bring all the samples to a final reaction volume to 100 μl, by adding BL binding buffer.
2. Incubate the binding reactions for 30 min at RT. Mix the tubes gently every 5 min.
3. Take 3 μl of [^{3}H]-BL microsome mixture into scintillation vials. This will serve to calculate total radioactive counts.
4. Separate the bound and free [^{3}H]-BL by vacuum filtering the mixture reaction through the presoaked (in 3.2.8) glass-fibre filter membrane (Whatman, GF/F) using a fritted glass microanalysis filter holder. To do this, keep the vacuum pressure constant, and use the filter forceps to place the glass-fibre membrane on the holder. Then, using a micropipette,

carefully place the binding reaction in the centre of the glass-fibre filter.

5. Wash the glass-fibre filter membrane by adding 2.5 ml of ice-cold BL-binding buffer in the fritted glass filter holder. Repeat the vacuum wash three times.
6. Use the filter forceps to carefully remove the glass-fibre filter membrane and place it immediately inside the scintillation tube containing 4 ml seintillation liquid. Do this for each sample until complete.
7. Bring the samples to the scintillation counter. Bound BL was quantified by scintillation counting as described in *(1)*.
8. Import the raw binding data into the Kaleida graph® software (Synergy) that allows calculation of binding affinity, dissociation constant (K_d) and graphic plotting of the results. Specific binding is calculated by subtracting background binding in the presence of 100-fold unlabelled BL from total binding without unlabelled BL.

3.3. Photoaffinity Crosslinking Assay for Microsomal Binding

3.3.1. Preparation of Microsomes

1. Grow 5-day-old etiolated seedlings. We use ~1,500 seeds (=40 ng) per experiment. Use wild-type Col-0 plants as a negative control.
2. Grind the fresh tissue using a mortar and pestle in 2 mL extraction buffer (50 mM Tris–HCl pH 7.5, 150 mM NaCl, 5 mM EDTA, 2 mM dithiothreitol and protease inhibitor cocktail [Sigma]). Transfer to Eppendorf tubes and keep the samples on ice.
3. Centrifuge the grinded tissues using a benchtop centrifuge at 2,800 rpm (10,000 × g) for 10 min at 4°C (a 20 μl aliquot can be saved for WB control, P3).
4. Ultracentrifuge the supernatant at 29,000 rpm (100,000 × g) for 30 min at 4°C.
5. Gently solubilize the microsomal pellet in extraction buffer containing 0.5% Triton X-100 (1 mg protein ml^{-1}) (a 20 μl aliquot can be saved for WB control, M4).
6. Quantify the microsomal concentration by Bradford assay according to manufacturer's instructions.

3.3.2. Immunoprecipia-tion of GFP Tagged Receptor

1. Use 1:1000 anti-GFP antibody and 20 μl of immobilized-protein A per ml of microsomal fraction.
2. Incubate IP for 16 h (overnight) with gentle rocking at 4°C.
3. Wash beads by centrifugation at 250 × g for 1 min three times in TBS buffer.
4. Transfer beads to an Eppendorf tube and give a final wash in BL-binding buffer.
5. Re-suspend the beads in 200 μl of BL-binding buffer and aliquot 100 μl/tube.

3.3.3. BPCS Binding Reaction

1. Using dim light in the dark room, add 1 μM BPCS compound and incubate with the immunoprecipitate for 20–30 min at room temperature.
2. Wash the beads with 1 ml of BL-binding buffer, by centrifugation 250 × g for 1 min for three times.
3. Suspend the resulting pellet in 250 μl of BL-binding buffer in a well of a 24-well plate (2.5 cm diameter).
4. Illuminate with UV lamp for 30 min on ice. Place the UV lamp 6 cm away from the samples.
5. After UV irradiation, harvest the reaction in Eppendorf tubes.
6. Take off the supernatant and re-suspend the pellet in 20 μl of SDS buffer.
7. Heat the samples for 2 min at 95°C. To remove the resin, centrifuge the samples one more time, and move the supernatant to a clean tube.
8. Separate the proteins of the supernatant by SDS-PAGE using 4–20% polyacrylamide gels.
9. Transfer the proteins to a nitrocellulose membrane by electroblotting.
10. Pre-incubate the membrane in blocking buffer for 30 min.
11. Incubate with biotin monoclonal antibody (Jackson ImmunoResearch) at a dilution of 1:3,000 in blocking buffer at room temperature for 2 h.
12. Rinse the membrane three times for 10 min each in T-TBS.
13. Secondary antibody incubation using a rabbit anti-mouse IgG secondary antibody conjugated to horseradish peroxidase (Bio-Rad) at a dilution of 1:5,000 in blocking buffer at room temperature for 2 h.
14. Wash the membrane three times for 10 min each with T-TBS.
15. Detect protein signal using SuperSignal Pico West chemiluminescent Substrate (Pierce) and BioMax film (Kodak).
16. After detection by chemiluminescence, use the same membrane for additional Western blotting using anti-GFP antibody (1:3,000) to detect BRI1-GFP fusion protein.
17. Incubate a goat anti-rabbit IgG conjugated to alkaline phosphatase (1:5,000) (BioRad) as a secondary antibody.
18. Develop the alkaline phosphatase reaction according to standard protocols.

4. Notes

1. [^{3}H]-BL-binding assays. Wash and dry carefully the tissue harvested for the membrane fraction preparation using

paper towels. Several parameters may affect the quality/ yield of membrane protein extraction, such as the tissue grinding quality and temperature at which the process take place. Make sure to keep the reagents at 4°C and pre-chill the centrifuges ahead of centrifugation and ultracentrifugation.

2. Tritium labelled BL. It is necessary to consider the safety rules when handling tritium labelled BL. Although the beta radiation from mCi quantities is not an external radiation hazard, tritium can enter the body by absorption through the skin, as well as ingestion and inhalation. Before starting a [^{3}H]-BL-binding experiment, read how to store and dispose radioactive waste in accordance with radiation safety requirements (http://hss.energy.gov/NuclearSafety/techstds/standard/hdbk1129/hdbk1129-07.pdf).
3. Immunoprecipitation time using GFP antibodies can last from 2 h to 16 h and still works fine. It can be adapted depending on the experiment schedule. However, if protein degradation occurs, shorter incubation time should be used.
4. Place the binding reaction in the centre of the membrane filters and pour immediately the washing buffer. Carefully handle membrane filters without damaging them using polished stainless steel forceps with unserrated tips.

References

1. Wang, Z. Y., Seto, H., Fujioka, S., Yoshida, S. and Chory, J. (2001) BRI1 is a critical component of a plasma-membrane receptor for plant steroids. *Nature* 410, 380–383.
2. Caño-Delgado, A., Yin, Y., Yu, C., Vafeados, D., Mora-García, S., Cheng, J. C., Nam, K. H., Li, J. and Chory, J. (2004) BRL1 and BRL3 are novel brassinosteroid receptors that function in vascular differentiation in Arabidopsis. *Development* 131, 5341–5351.
3. Kinoshita, T., Caño-Delgado, A., Seto, H., Hiranuma, S., Fujioka, S., Yoshida, S. and Chory, J. (2005) Binding of brassinosteroids to the extracellular domain of plant receptor kinase BRI1. *Nature* 433, 167–171.
4. Clouse, S. D. (2002) Arabidopsis mutants reveal multiple roles for sterols in plant development. *Plant Cell* 14, 1995–2000.
5. Belkhadir, Y. and Chory, J. (2006) Brassinosteroid signaling: a paradigm for steroid hormone signaling from the cell surface. *Science* 314, 1410–1411.
6. Li, J. and Chory, J. (1997) A putative leucine-rich repeat receptor kinase involved in brassinosteroid signal transduction. *Cell* 90, 929–938.
7. Friedrichsen, D. M., Joazeiro, C. A., Li, J., Hunter, T. and Chory, J. (2000) Brassinosteroid-insensitive-1 is a ubiquitously expressed leucine-rich repeat receptor serine/threonine kinase. *Plant Physiol* 123, 1247–1256.
8. Seto, H.E.A. (1998) A general approach to synthesis of labeled brassinosteroids: preparation of [25,26,27-^{2}H7]brassinolide with 60% isotopic purity from the parent brassinolide. *Tetrahedr Lett* 39, 7525–7528.
9. He, Z., Wang, Z. Y., Li, J., Zhu, Q., Lamb, C., Ronald, P. and Chory, J. (2000) Perception of brassinosteroids by the extracellular domain of the receptor kinase BRI1. *Science* 288, 2360–2363.

Chapter 8

Binding Assays for Abscisic Acid Receptors

Fawzi A. Razem and Robert D. Hill

Abstract

The plant hormone abscisic acid (ABA) is a major growth regulator mediating several aspects of plant development and stress-induced processes. For the hormone to function within the cell, it must first interact with a proteinaceous receptor to commence transmission of a signal to downstream components that elicit a cellular response. With the identification of several ABA receptors, there is an increased interest in receptor–ABA interaction studies and a demand for easy-to-follow protocols to allow measurement and quantification of ABA binding. Here, we present protocols for radioligand-binding measurements with particular focus on the ABA-binding of the recently identified ABA receptors, ABAP1 and FCA.

Key words: Affinity, ABA-binding assays, association rate constant, competition assays, dissociation rate constant, equilibrium dissociation constant, radioligand-binding assays, saturation kinetics, ^{3}H-(+)-ABA.

1. Introduction

The radioligand-binding assays are based on the incubation of a receptor (R), either as a purified protein or within a cellular fraction or tissue, with a radiolabeled compound, here referred to as ligand (L), and the ability to separate the unbound free ligand, thus allowing the quantification of the bound labeled ligand *(1, 2)*. Receptor-hormone binding assays are commonly used to quantify several binding parameters and determine the affinity of the hormone for its receptor. From a pharmacological perspective, receptor–ligand interaction follows the Law of Mass Action, in which the ligand binds reversibly to a single binding site:

$$R + L \underset{k_{off}}{\overset{k_{on}}{\rightleftarrows}} RL \qquad (8.1)$$

Sean Cutler, Dario Bonetta (eds.), *Plant Hormones: Methods and Protocols, Second Edition, vol. 495*

DOI 10.1007/978-1-59745-477-3_8 Springerprotocols.com

When the receptor, R, is physically in contact with its ligand, L, interaction to form the RL complex occurs at a rate determined by the association rate constant, k_{on}, and the concentrations of R and L. This interaction is reversible, with the rate of breakdown of the RL complex being determined by the dissociation rate constant, k_{off}, and the concentration of RL. Binding equilibrium is reached when the rates of formation and dissociation are equal, i.e.,

$$[R_{eq}][L_{eq}]k_{on} = [RL_{eq}]k_{off} \quad (8.2)$$

By rearranging equation (8.2), it is possible to calculate the equilibrium dissociation constant K_d, a critical parameter in receptor–ligand binding kinetics:

$$K_d = k_{off}/k_{on} = [R_{eq}][L_{eq}]/[RL_{eq}] \quad (8.3)$$

$[R_{eq}]$, $[L_{eq}]$, and $[RL_{eq}]$ are the equilibrium concentrations of R, L, and RL, respectively.

The dissociation constant, K_d, provides a measure of how tightly R binds L and can be interpreted to represent the equilibrium concentration of ligand that results in 50% occupancy of the receptor, i.e., a concentration where ligand is bound to half of the receptor binding sites. The lower the K_d value the higher the affinity of the receptor for its ligand.

There are several components of measuring radioligand-binding that require consideration to properly assess ligand–receptor characteristics. Nonspecific binding can occur as a result of radioligand binding to non-saturable components, such as lipids. Nonspecific binding is assessed by adding a large excess of non-radiolabeled ligand to the mixture containing the radiolabeled ligand/receptor, measuring the activity remaining with the receptor. Specific binding is determined by subtracting this value from the total radioactivity associated with the receptor without the addition of unlabeled ligand:

$$\text{Specific binding} = \text{total binding - nonspecific binding} \quad (8.4)$$

Saturation binding experiments use a fixed amount of receptor in the presence of a range of ligand concentrations (**Fig. 8.1**A). This information can be used to determine the rate constants for the ligand/receptor couple. Thus, transformation of the saturation data to a Scatchard plot (**Fig. 8.1B**) allows the determination of several parameters, such as maximum binding (B_{max}) and K_d. If the binding follows the Law of Mass Action, then K_d obtained from the ratio k_{off}/k_{on} (Equation 8.3) should agree with saturation analysis.

Competition assays, on the other hand, are used to determine the degree of specificity associated with a receptor binding site. With this type of assay, it is possible, for example, to determine whether an enantiomer of the native ligand also has the potential

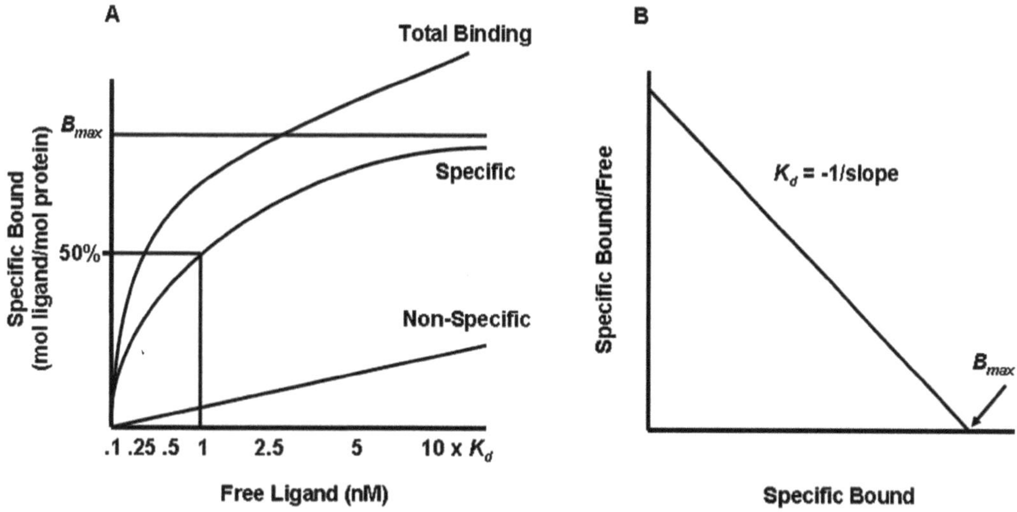

Fig. 8.1. Saturation and Scatchard analyses. (**A**) Saturation results can be plotted with specific bound ligand on the *Y*-axis and free unbound ligand on the *X*-axis. Curves for specific, non-specific, and total binding are shown for comparison. The concentrations of free ligand added are 0.1, 0.25, 0.5, 5, and 10X the receptor's K_d. (**B**) The results of saturation experiments can be plotted with specific bound/free ligand on the *Y*-axis and specific bound on the *X*-axis to obtain a Scatchard plot. The K_d value is –1/slope and the B_{max} values is the X-intercept.

to bind to the receptor site or if modification of a key residue in the ligand, i.e., substitution of a carboxyl group with an alcohol moiety, alters binding. The advantage of the competition experiment is that the assay can be performed without requiring labeling of the competing molecule. In a competition experiment, increasing concentrations of unlabeled, competing ligand is added in the presence of a fixed concentration of the radiolabeled ligand and the receptor. If the unlabeled ligand competes for the binding site then, as the concentration of the unlabeled ligand increases, the amount of labeled ligand bound to the receptor decreases.

2. Materials

1. Binding buffer: 25 mM Tris–HCl buffer, pH 7.3, 250 mM sucrose, 5 mM $MgCl_2$, and 1 mM $CaCl_2$. Buffer is stored at 4°C.
2. Purified recombinant protein, such as ABAP1 or FCA. Purity of the protein should be determined on SDS-PAGE on the day of use.
3. Labeled ABA stock (e.g., 10 μM ^3H-(+)-ABA). ABA is first dissolved in methanol and subsequent working stocks are

dissolved in water and should be stored at −20°C. The ^{3}H-(+)-ABA used should have high specific activity (*see* **Note 1**).

4. Unlabelled (+)-ABA stock of 1 mM. For competition assays, similar stocks from (−)-ABA, trans-ABA can also be prepared.
5. Dextran T70-coated charcoal (DCC), Sigma.
6. Nitrocellulose membranes or any other filters with low affinity to ABA.

3. Methods

3.1. ABA-Binding Assay

The specific activity of the radiolabeled ligand that should be used in the assays depends upon the amount of receptor protein in the mixture and the affinity of the receptor for the ligand (*see* **Note 1**). It may be necessary to compromise on the specific activity of the ligand based on the amount of receptor and radiolabeled ligand available to the investigator. It is necessary to have sufficient radioactivity, disintegrations per minute (dpm), with the bound ligand to obtain plots that provide an accurate estimate of K_d (*see* **Note 2**). The concentration of radioligand is best determined based on the K_d of the receptor. For a potential receptor with unknown binding characteristics, preliminary experiments using K_d values of known receptors can be used. Alternatively, using concentrations in the lower range in which the hormone is known to be effective in the tissue can be attempted. Once having established an approximate K_d, a second set of experiments using concentrations based on this approximate K_d can be used. Normally in initial binding experiments, the concentration should not be less than 0.8 K_d, with radioactivity in the range of 100–1000 dpm above background, associated with the receptor. The proportion of the receptor bound to ligand at equilibrium can be determined based on a rearrangement of equation (8.3) as follows:

$$K_d = [R_{eq}][L_{eq}]/[RL_{eq}]$$

$[R_{eq}] = [R_t] - [RL_{eq}]$; where $[R_t]$ = initial receptor concentration.

Therefore,

$$K_d = ([R_t] - RL_{eq}])[L_{eq}]/[RL_{eq}]$$

and

$[RL_{eq}]/([R_t] - [RL_{eq}]) = [L_{eq}]/K_d$ and, dividing top and bottom by $[R_t]$ to convert to proportional binding,

$F_{bound}/(1\text{-}F_{bound}) = [L_{eq}]/K_d$, where F_{bound} = fraction of receptor bound to ligand at equilibrium. In most instances, $[L_{eq}]$ is a good approximation of the initial ligand concentration.

The four ABA receptors identified have strikingly similar K_d ranging from 19 nM to 32 nM ABA *(3–6)*. ABAP1 has a K_d of 28 nM. Thus, for ABAP1, at a ligand concentration of $0.8 \times K_d$, the ratio of bound ligand to free ligand at equilibrium is 1.25, indicating that 55.6% of the receptor is bound to ligand. A series on increasing concentrations between $0.8 \times K_d$ to $10 \times K_d$ (90% of the receptor bound) should provide sufficient data points for saturation binding plots. Note that a 12-fold increase in ligand concentration is required to achieve an approximately 80% increase in the amount of ligand bound to receptor. It is, thus, very important to maintain the specific activity of the radiolabeled ligand as the concentration of ligand increases. For ABAP1, 1 μg of ABAP1 (52 kDa) should bind approximately 0.019 nmol ABA (e.g., at a mol to mol ratio) at saturation in a 100 μl binding assay. The specific activity of the radiolabeled ABA should, therefore, be in the range of 5000–50,000 dpm/ nmol ABA, in order to find 100–1000 dpm bound to 1 μg ABAP1 at saturation. At 55% saturation ($0.8 \times K_d$), the specific activity would need to be almost double the above amounts.

1. Two sets of 1.5 ml Eppendorf tubes are prepared. One set is used to determine total binding and the second set is used to determine the nonspecific binding of ^{3}H-(+)-ABA. All assays are carried out in 100 μl assay volumes and in triplicate.
2. Add 1 μl of a 10 μM ^{3}H-(+)-ABA stock to 98 μl binding buffer. If the concentration of ^{3}H-(+)-ABA stock is different, the volumes can be adjusted accordingly. The final concentration of ^{3}H-(+)-ABA in the binding mixture therefore should be several times higher than K_d and will allow equilibrium to be reached. For initial binding assays and all other assays that do not require equilibrium, a lower concentration of ^{3}H-(+)-ABA in the binding mixture, e.g., $0.8 \times K_d$ to $2 \times K_d$ should be used.
3. For the determination of total binding, add 1 μl of ABAP1 (from a 1 μg/1 μl solution) to each tube to make a final volume of 100 μl binding assay in the first set expect first three tubes just containing 99 μl buffer and 1 μl ^{3}H-(+)-ABA stock. The first three tubes of this set will contain only buffer and labeled ligand, but no protein and will be used to determine the efficiency of unbound ligand separation (see below). The amount of protein in the binding mixture can vary to enhance radioligand binding.
4. For the determination of nonspecific binding, a second set of tubes are prepared as above except that a 50–100 fold higher (than ^{3}H-(+)-ABA) concentration of unlabeled (+)-ABA is added in each tube containing protein.

5. Mix by tapping, spin gently, and place tubes at 4°C for 60 min to reach steady-state conditions (*see* **Note 3**).

3.1.1. Separation of Bound ^{3}H-(+)-ABA from Free Ligand

Following 1 h incubation, the bound ^{3}H-(+)-ABA is separated from the free unbound ABA by two methods, filtration and/or adsorption. This is a critical step and there are few things that should be kept in mind; the most important is to keep this process as quick as possible to minimize the dissociation of bound ligand. The filtration is based on trapping the bound ^{3}H-(+)-ABA on a filter (e.g., a nitrocellulose membrane) while allowing the free ligand to pass through, whereas the adsorption is based on trapping the free ligand by adsorbing resin (e.g., dextran coated charcoal, DCC) and leaving the receptor-ligand complex into solution.

1. On a filter paper (0.5 × 0.5 cm), pipette the binding mixture and allow the mixture to pass through. This can be facilitated by a low-pressure water aspirator.
2. Pipette 3 × 0.5 ml 10% binding buffer to wash the free ligand, dry and place the filter into a scintillation vial (*see* **Notes 4** and **5**).
3. For adsorption, add 0.6% DCC in each tube, mix by tapping on the tube and centrifuge at 1330 rpm for 5 min. Immediately transfer the supernatant into a scintillation vial. It is important to determine the amount needed to adsorb the free ABA, particularly in tubes containing unlabeled ABA for nonspecific determination (*see* **Note 6**).

3.2. Competition Assays

The competition experiments, often known as inhibition analysis, allow the determination of specific binding by subtracting nonspecific binding from the total binding of a ligand and also can be used to determine the affinity of structurally related unlabeled ligands for a receptor binding site. In ABA competition assays, various concentrations of unlabeled ligands (e.g., (+)-ABA, (−)-ABA, trans-ABA) compete with a fixed concentration of ^{3}H-(+)-ABA for binding to the receptor. If these unlabeled ligands compete for the receptor's binding site, then as their concentration increases in the binding assay, the binding of ^{3}H-(+)-ABA decreases. The concentration of the ^{3}H-(+)-ABA should be close to the receptor K_d. If a high concentration of ^{3}H-(+)-ABA is used, then high concentrations of the unlabeled ligand are required to compete with the ^{3}H-(+)-ABA binding and subsequently this may cause problems during the separation of the unbound ligand. Five to ten different concentrations of unlabeled ligand are sufficient to define an inhibition curve for a fixed ^{3}H-(+)-ABA concentration (e.g., 0.8 × K_d). The data are usually plotted on a log rather than linear plot of ABA concentrations and therefore it is helpful to use a series of fold dilutions (e.g., 0, 10, 30, and 100-fold of K_d).

1. Prepare tubes as described in **Section 3.1**, except that the final ^{3}H-(+)-ABA concentration in all tubes is around 0.8 × K_d

(e.g., 22 nM for ABAP1 and 15 nM ^{3}H-(+)-ABA for FCA). Receptor protein (e.g., ABAP1) is not added at this stage.

2. To each tube, add a series of unlabelled (+)-ABA concentrations equal to 0, 1, 10, 30 , 50, 100, 200 × ^{3}H-(+)-ABA available in the binding mixture.
3. Add receptor protein and mix gently by tapping on the tube.
4. Allow steady-state conditions to be reached (approximately 60 min at 4°C).
5. Add DCC in each tube, with amount pre-determined to adsorb the highest concentration of all free ABA added (*see* **Note 6**).
6. Proceed as described in **Section 3.1** for counting.

3.3. Equilibrium Analysis

Equilibrium and steady-state conditions can be determined by the association and dissociation assays. To measure the rate of radioligand association, receptors are incubated with a certain concentration of ligand and binding is measured at different times. The dissociation rate is determined by incubating the receptor with the radioligand, usually, though not necessarily, until steady-state conditions are reached. A high concentration of unlabeled ligand is then added and the decrease in radioactivity associated with the receptor measured as a function of time.

Prepare binding assays as described in **Section 3.1**.

1. Add DCC in three tubes immediately after starting the binding reaction, e.g., after adding the receptor protein. This will be *time 0*.
2. Add DCC thereafter in three additional tubes every 5–10 min and process for separating the bound from the unbound as described in **Section 3.1.1** until you collect samples representing a time range of 90 min.
3. Count the amount of ^{3}H-(+)-ABA bound and determine the moles of ligand per mole of receptor at each time point. Plot specific binding on the *y*-axis and time on the *x*-axis to determine $t_{1/2}$, the time point where half of the binding sites are occupied (**Fig. 8.2**A).
4. Estimate maximum binding B_{est} of specific binding. This value should be close to that calculated by the saturation studies (i.e., B_{max}) (*see* **Section 3.4**). Plot $\ln(B_{est}/(B_{est} - B_t)$ against time (**Fig. 8.2B**), where B_t is the specific binding at time *t*. The slope of the straight line is the estimated association rate constant (k_{est}).
5. Calculate the association constant, k_{on} as shown in **Fig. 8.2B**.
6. For the dissociation rate (k_{off}) determination, prepare binding assays as described in **Section 3.1** and allow the binding to proceed for 60 min.

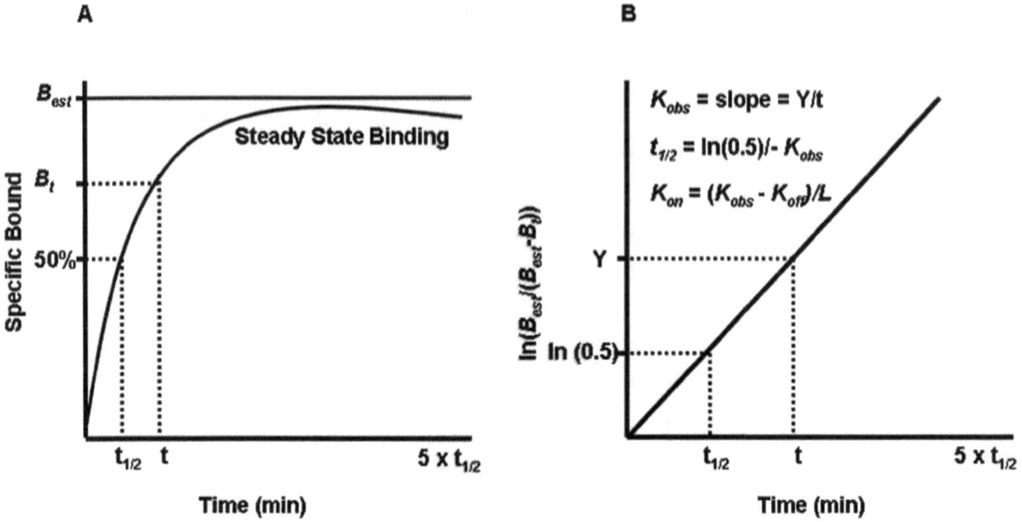

Fig. 8.2. Receptor-ligand association kinetics. (**A**) Time course of ligand binding can be monitored by plotting specific bound ligand on the *Y*-axis vs time at which samples were collected on the *X*-axis (*see* **Section 3.3**). Steady state conditions are reached when no increase in ligand binding is observed and this point is the maximum (estimated) binding B_{est}. Time at which $\frac{1}{2}B_{est}$ is reached is the $t_{1/2}$. Curve of ligand association can be defined over a time period equal to approximately $5 \times t_{1/2}$. B_t is the specific binding at time *t*. (**B**) The rate of association, K_{on}, can be calculated by transforming results from the association experiment (**A**) in which the natural logarithm (ln) of binding (i.e., $\ln(B_{est}/(B_{est}-B_t))$) is plotted on the *Y*-axis vs time on the *X*-axis. To calculate K_{on}, the observed association K_{obs} (i.e., slope) should be first determined. Half-time association ($t_{1/2}$) is then determined by finding $\ln(0.5)/-K_{obs}$ and K_{on} is subsequently determined by subtracting K_{obs} from K_{off} over the ligand concentration (*L*) used in the association experiment (i.e., $(K_{obs}- K_{off})/L$) as shown.

7. Add at least 100-fold higher concentration of unlabeled (+)-ABA than the concentration of ^{3}H-(+)-ABA present in the binding mixture. This will allow exchange of the ^{3}H-(+)-ABA with non-labeled ABA at a rate determined by the k_{off}, allowing calculation of the dissociation rate of the receptor–ligand complex.
8. Collect samples every 15–30 min, including the time when the unlabeled (+)-ABA was added (B_0, binding before dissociation started) (**Fig. 8.3**A). Immediately add DCC and isolate bound from unbound as described in **Section 3.1.1.**
9. Calculate B_0 of specific binding (Fig. **8.3A**). This value should be close to B_{max} calculated by the saturation studies (*see* **Section 3.4**). Plot $\ln(B_t/B_0)$ against time (**Fig. 8.3B**), where B_t is the specific binding remaining at time *t* and B_0 is binding before dissociation started. The negative slope of the straight line is the dissociation rate constant (k_{off}).

3.4. Saturation Analysis

Two parameters can be studied from saturation experiments, maximum binding B_{max} and equilibrium dissociation constant K_d. The equilibrium dissociation constant K_d represents the

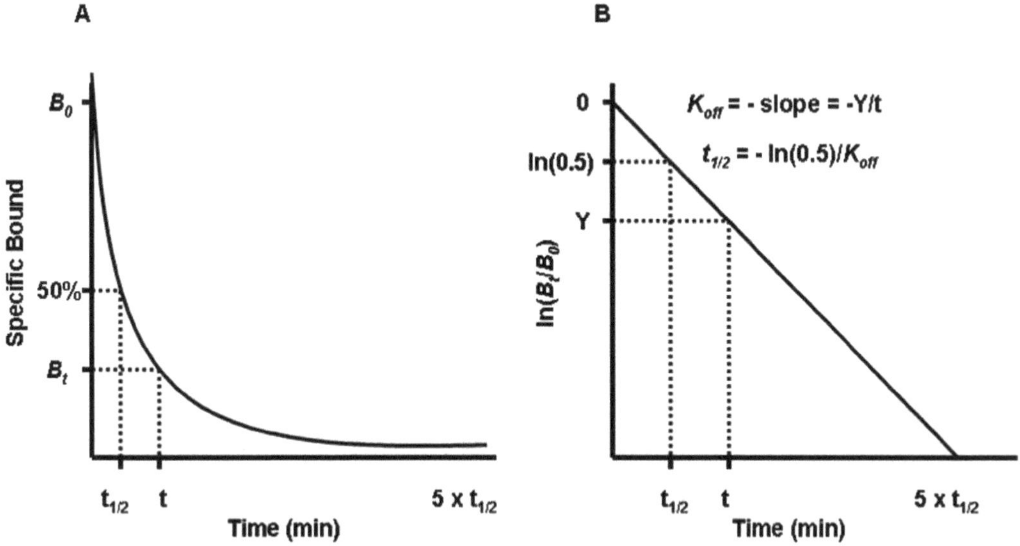

Fig. 8.3. Receptor–ligand dissociation kinetics. (**A**) Time course dissociation of ligand binding can be monitored by plotting the remaining of specific bound ligand (after the addition of high concentration of unlabeled ligand) on the *Y*-axis vs time at which samples were collected on the *X*-axis (*see* **Section 3.3**). Unlabeled ligand is added after steady state conditions are reached and binding before dissociation is started is B_0. Time at which ½B_0 is reached is the $t_{1/2}$. Curve of ligand binding dissociation can be defined over a time period equal to approximately $5 \times t_{1/2}$. B_t is the specific bound ligand remaining at time *t*. (**B**) Results from the time course association experiment (**A**) can be transformed to calculate the rate of dissociation, K_{off}, by plotting the natural logarithm (ln) of binding at time *t* over binding before dissociation started (i.e., $\ln(B_t/B_0)$) on the *Y*-axis vs time on the *X*-axis. The negative slope represents the K_{off} and the half-time dissociation ($t_{1/2}$) is determined by finding $-\ln(0.5)/K_{off}$ as shown.

concentration when half of the receptor binding sites are occupied with binding ligand (**Fig. 8.1A**). As the concentration of radioligand increases, a point is reached (B_{max}) where binding approaches saturation (**Fig. 8.1A**). A typical experiment contains six assays with three ^{3}H-(+)-ABA concentrations below K_d and three above K_d (**Fig. 8.1A**). The assays are set as above (**Section 3.1**), but with serial concentrations of ^{3}H-(+)-ABA (e.g., 0.1; 0.25, 0.5, 1, 2.5, 5, 10 × K_d) until steady-state conditions are reached. Two sets of tubes at each concentration, first set contains binding buffer, protein, and serial concentrations of ^{3}H-(+)-ABA to be used for total binding and a second set containing the same quantities of the above components (i.e., buffer, protein and ^{3}H-(+)-ABA) in addition to unlabeled (+)-ABA and this set is used for specific binding determination. The addition of higher concentrations of (+)-ABA (*see* **Section 3.1**) will competitively inhibit the binding of ^{3}H-(+)-ABA and thus allows the determination of nonspecific binding as mentioned in equation (8.4).

3.5. Data Analysis

A small, known aliquot of the reaction mixture can be counted and, knowing the molar concentration of the ligand in the mixture, the ligand specific activity can be determined. Most scintillation counters automatically correct for counting efficiency, so that the specific activity can be expressed as dpm per mole of ligand. The specific binding activity, as determined in equation (8.4) and representing the activity bound to the receptor, divided by the ligand specific activity yields the moles of ligand bound to the receptor. The molar concentration of the receptor is derived from the receptor protein concentration and its molecular size. These values can be used with the appropriate graphs to obtain the various values.

4. Notes

1. The radiolabeled ligand used for binding assays should have high specific radioactivity, generally 30 Ci/mmol or higher.
2. The concentration of bound ligand must provide adequate signal. Activities less than a 150 dpm may be insufficient to give accurate measurements, particularly if background levels are high. Generally, assays of radioactivity are taken to 1% statistical accuracy, i.e., 10,000 radioactive decay events are counted. Thus, for a sample containing 100 dpm, in a scintillation counter with 50% efficiency, this would require a counting time of 100 min. Low sample activities can be partially overcome by increasing the receptor concentration and specific radioactivity of the ligand. If low specific activity of the ligand remains a problem, iodinated radioligands can be considered *(2)*.
3. The assay mixture can be incubated at either room temperature or 4°C. Incubation at cold temperature increases the stability of the receptor, but may result in slower association. Association/dissociation rate constants can be determined as described in **Section 3.3.**
4. Free ^{3}H-(+)-ABA may bind to the filters, thus increasing the background counts. This can be reduced by presoaking the filters in a solution containing unlabeled (+)-ABA, membranes then washed and dried before use. Generally, filters should not be used for low affinity receptors (e.g., $k_d > 10^{-8}$ M) *(2)*.
5. It is very important to dry the filters following washing as a wet filter may interfere in counting. For tritiated ligand, such as ^{3}H-(+)-ABA, a wet filter can only be used if it is immersed in scintillation fluid for sufficient time (e.g., overnight) and shaken vigorously to dissolve the water before counting.
6. It is important to determine the amount of DCC needed to adsorb the highest ABA concentration added in the binding mixture. DCC does not discriminate between labeled and unlabeled ABA and if DCC is used in competition assays,

there has to be enough of it to adsorb all free ABA, both labeled or unlabeled. This can be determined by preparing a series of tubes containing binding buffer and ^{3}H-(+)-ABA concentration representing the highest ABA concentration combining labeled and unlabeled ABA used in any experiment plus a series of DCC resin. The amount of DCC that gives counts identical to the blank, i.e., readings without ^{3}H-(+)-ABA should be used.

References

1. Bylund, D.B., and Toews, M.L. (1993) Radioligand-binding methods: practical guide and tips. *Amer. J. Physiol.* 265, L421–L429.
2. Lazareno, S. (2001) Quantification of receptor interactions using binding methods. *J. Rec. Sig. Trans. Res.* 21, 139–165.
3. Razem, F.A., Luo, M., Liu, J.-H., Abrams, S.R., and Hill, R.D. (2004) Purification and characterization of a barley aleurone abscisic acid-binding protein. *J. Biol. Chem.* 279, 9922–9929.
4. Razem, F.A., El Kereamy, A., Abrams, S.R., and Hill, R.D. (2006) The RNA-binding protein FCA is an abscisic acid receptor. *Nature* 439, 290–294.
5. Liu, X., Yue, Y., Li, B., Nie, Y., Li, W., Wu, W.-H., and Ma, L. (2007) A G protein-coupled receptor is a plasma membrane receptor for the plant hormone abscisic acid. *Science* 315, 1712–1716.
6. Shen, Y.-Y., Wang, X.-F., Wu, F.-Q., Du, S.-Y., Cao, Z., Shang, Y., Wang, X.-L., Peng, C.-C., Yu, X.-C., Zhu, S.-Y., Fan, R.-C., Xu, Y.-H., and Zhang, D.-P. (2006) The Mg-chelatase H subunit is an abscisic acid receptor. *Nature* 443, 823–826.

Chapter 9

Cytokinin Sensing Systems Using Microorganisms

Masayuki Higuchi, Tatsuo Kakimoto, and Takeshi Mizuno

Abstract

The cytokinin class of plant hormones is perceived by transmembrane His-kinases (His-kinases) of the two-component system, otherwise known as the His-Asp phosphorelay system. When cytokinin receptors perceive cytokinins, they are autophosphorylated at a conserved His residue. The phosphoryl group is then transferred to downstream components of the His-Asp phosphorelay system. When the gene for a cytokinin receptor is introduced into yeast or *Escherichia coli*, the corresponding receptor feeds the phosphoryl group to the phosphorelay system of the host, in a cytokinin-dependent manner. Therefore, these microorganisms can be used as convenient cytokinin sensors, and can also be used to understand the properties of cytokinin-receptors. Furthermore, they may be used to screen for cytokinin agonists and antagonists, which would potentially be useful to regulate the growth of crops.

Key words: Cytokinin, Sensor, Phosphorelay, Two-component system, Histidine kinase, Phosphotransfer, Cytokinin receptor, CRE1, AHK4, AHK2, *Saccharomyces cerevisiae*, *Escherichia coli*, β-Galactosidase.

1. Introduction

Cytokinins play important roles in cell division, differentiation, allocation of nutrients, inhibition of senescence, activation of axillary meristems, and inhibition of cell elongation (for a review, see *(1)*). Natural cytokinins are adenine derivatives with a dimethylallyl (isopentenyl) side chain at the N^6 position. Although many cytokinin derivatives are known, until recently it was impossible to prove which cytokinin molecular species were active because many of these were metabolized to different molecular species when administered to plants *(2, 3)*. This question is being solved using a cytokinin-sensing system described in this chapter and in vitro receptor-binding assays described elsewhere *(4, 5)*.

Sean Cutler, Dario Bonetta (eds.), *Plant Hormones: Methods and Protocols, Second Edition, vol. 495*

DOI 10.1007/978-1-59745-477-3_9 Springerprotocols.com

His-kinases are receptor components of the His-Asp phosphorelay systems in bacteria, fungi, and plants. A His-Asp phosphorelay system typically consists of a His-kinase(s) and a response regulator(s). In the presence of a signal, the His-kinase is autophosphorylated at the conserved His, and then the phosphoryl group is directly or indirectly transferred to the conserved Asp of a response regulator, modulating the activity of an attached signaling domain (in many cases, with a function in transcriptional regulation). An intermediate component(s), HPt domain protein(s), may also be involved in the phosphorelay **(Fig. 9.1)** *(6)*.

The cytokinin receptor CRE1 (also known as AHK4 or WOL) was independently identified as the causal gene of a cytokinin-insensitive mutant *(7)*, or from database mining and functional analyses *(8)*. In these studies, the *Arabidopsis CRE1/AHK4* was expressed in *Saccharomyces cerevisiae (7)*, *Schizosaccharomyces pombe (8)* or *E. coli (8)* to prove that CRE1/AHK4 was indeed a cytokinin receptor. For example, when the *CRE1/AHK4* gene was expressed in either yeast or *E. coli*, the phosphorelay from CRE1/AHK4 to the host His-Asp phosphorelay system occurred in a cytokinin-dependent manner *(7–9)*. In addition, proof that CRE1/AHK4 is a His-kinase that is activated by cytokinins has been demonstrated by the binding of cytokinins to membrane-fractions of *S. pombe (5)* or *E. coli (10)* or to whole *E. coli* expressing *CRE1/AHK4 (10)*. Furthermore, binding of CRE1/AHK4 by cytokinins has also been shown by CRE1/AHK4 activation in vitro *(9)*. In this chapter, we describe detailed methods of a cytokinin-sensing systems using *S. cerevisiae* and *E. coli*.

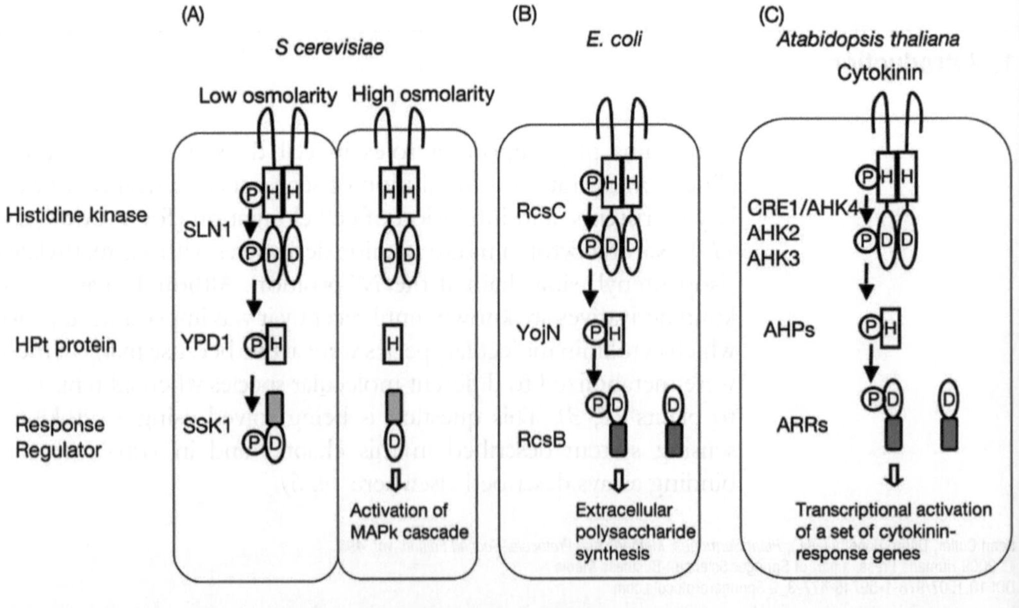

Fig. 9.1. Several examples of phosphorelay signaling. **(A)** Osmo-sensing in *S. cerevisiae*. **(B)** *E. coli* regulatory systems for extracellular polysaccharide synthesis. **(C)** Cytokinin-perception system in *Arabidopsis*.

2. Materials

2.1. *Saccharomyces cerevisiae* Strains

1. TM182 : *MATa leu2 ura3 his3 sln1::hisG* [pSSP25] (*see* **Notes 1** and **2**) *(11)*: A yeast strain that lacks *SLN1*, which is the sole gene for a His-kinase in *S. cerevisiae*. pSSP25 carries the *PTP2* gene under the control of the *GAL1* promoter, and in the presence of galactose, suppresses the lethality caused by the *sln1* mutation. pSSP25 has the *URA3* selection marker.
2. TM182 carrying either of the following plasmids (*see* **Notes 1** and **2**):
   p415CYC1-CRE1
   pHM1-AHK2
   pHM1-AHK2ΔN (see below)

All of these plasmids carry the ampicillin-resistance gene that is functional in *E. coli* and the *LEU2* selection-marker gene that is functional in yeast. A gene for a cytokinin receptor (CRE1; AHK2; or AHK2ΔN, a modified AHK2) is under the control of the truncated weak *CYC1* promoter *(12)*.

2.2. Yeast Media for Cytokinin Assay

1. Phytohormone stock solution: Cytokinins (*trans*-zeatin, *cis*-zeatin, isopentenyladenine and any other samples to be tested). Dissolve samples in dimethylsulfoxide (DMSO) at 5000 times the final concentrations of those in yeast assay media.
2. -ura/-leu/galactose medium and -ura/-leu/glucose medium containing cytokinins:

6.7 g/l Difco™ Yeast Nitrogen base w/o amino acid (Becton Dickinson, Meylan, France), 0.67 g/l -ura/-leu DO Supplement (Clontech, Cambridge, UK), 20 g/l galactose or glucose, 15 g/l Agar (Nacalai Tesque, Kyoto, Japan) (for plates only). Adjust the pH to 5.8. After autoclaving at 121°C for 20 min, add cytokinin stock solution to make cytokinin-containing plates with the appropriate cytokinin concentrations.

2.3. *E. coli* Strains

1. A unique *E. coli* mutant strain is employed, which is named KMI001 [*cps::lacZ rcsC ara thi* Δ(*pro-lac*) Δ(*wzc-wca*)], carrying the *lacZ* reporter gene.
2. The *lacZ* gene, encoding β-galactosidase, is integrated into the chromosome and is under the control of the *cps* promoter, which is regulated through the RcsC-YojN-RcsB phosphorelay signal transduction in response to an as yet unidentified environmental signal in *E. coli*.
3. In the RCS-phosphorelay system, the *rcsC* gene encodes a His-kinase. Note that the strain KMI001 lacks the *rcsC* gene.

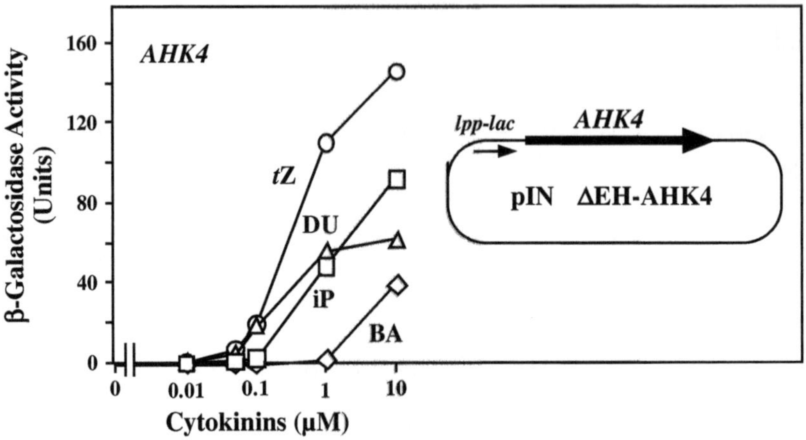

Fig. 9.2. Quantitaive measurement of the CRE1/AHK4 cytokinin receptor activity in *E. coli*. The *CRE1/AHK4* gene (or cDNA) was cloned onto an *E. coli* expression vector, as schematically illustrated. The plasmid was transferred into the *E. coli* strain KMI001 (Δ*rcsC cps::lacZ*). These transformants were grown overnight at 25̊C in LB, supplemented with and without the indicated concentrations of cytokinins. Then, β-galactosidase activity was assayed. Cytokinins tested were *trans*-zeatin (*tZ*), isopentenladenine (iP), benzyl-adenine (BA), and diphenylurea (DU).

4. KMI001 carrying pINΔEH-AHK4, which express the *CRE1/AHK4* cDNA in *E coli*. (*see* **Fig. 9.2**). This plasmid confers ampicillin-resistance to *E.coli*.

2.4. Other

1. Luria-Bertani (LB) medium: 10 g Bacto™ Tryptone (DIFCO), 5 g Bacto™ Yeast Extract (DIFCO), 5 g NaCl in 1 L of H2O. After being autoclaved, the LB medium is supplemented with glucose and sodium phosphate buffer (pH 7.0) to final concentrations of 40 and 50 mM, respectively. If required, ampicillin is added to a final concentration of 50 μg/mL. To prepare solid agar-plates, 1.5% agar (w/v) is added to the medium before autoclaving.
2. Z-buffer: 4.8 g NaH_2PO_4, 21.48 g Na_2HPO_4 ($12H_2O$), 0.75 g KCl, 0.246 g $MgSO_4$ ($7H_2O$) in 1 L of H_2O. Just before using this buffer, β-mercaptoethanol (3.5 μL/mL) is added. ONPG (o-nitrophenyl-β-galactopyranosde) (4 mg/mL) is dissolved in 100 mM sodium-phosphate buffer (pH 7.0). If required, X-Gal (5-bromo-4-chloro-3-indolyl-β-D-galastoside) is added to agar-plates at 40 μg/mL.

3. Methods

Cytokinin receptors CRE1/AHK4, AHK2 and AHK3 are membrane-spanning receptors that carry an extracellular cytokinin-sensing domain and intracellular His-kinase and receiver domains

(1). In response to cytokinins, these receptors are activated to autophosphorylate, the conserved His residue in the histidine kinase domain. The phosphoryl group is then transferred to the conserved Asp residue in the receiver domain in the same molecule, and then transferred to the conserved His residue in an HPt domain protein before ultimately being transferred the conserved Asp of a response regulator *(1)*. Similar His-Asp phosphorelay systems are present in yeast and *E. coli*, which can be exploited to make cytokinin-sensing systems **(Fig. 9.1)** *(7, 8)*.

In the yeast *S. cerevisiae*, there is a continuous phosphoflow from SLN1 (His-kinase) to YPD1 (HPt domain protein) and to SSK1 (response regulator) at non-stress osmolarities. In response to hyperosmolarity, the phosphorelay ceases, leading to dephosphorylation of SSK1. Dephosphorylated SSK1 activates a mitogen-activated protein kinase (MAPK) cascade, which in turn induces osmolarity responses. Disruption of SLN1 is lethal, because dephosphorylated SSK1 over-activates the MAPK cascade. The lethality can be suppressed by expressing *PTP2* (a gene for a phosphatase that acts on HOG1 MAPK); *PPT2* expression can be under the control of the galactose-inducible *GAL1* promoter *(11, 13)*. When a cytokinin receptor gene *CRE1/AHK4* is expressed in the TM182 strain, the yeast will grow in the absence of galactose if cytokinin is present because phosphoryl group is transferred from CRE1/AHK4 to yeast YPD1 *(7)*.

Arabidopsis has two other cytokinin receptors, AHK2 and AHK3. Growth of *sln1Δ* yeast (TM182) carrying either *AHK2* **(Fig. 9.3)** or *AHK3* (data not shown) is also accelerated by lower levels of cytokinins than those required for activation of CRE1/AHK4. However, yeast expressing AHK2 **(Fig. 9.3)** or AHK3 (data not shown) grows even in the absence of cytokinins, albeit slower than in their presence. This suggests that AHK2 and AHK3 have some basal activity in yeast.

CRE1/AHK4 is predicted to have two transmembrane segments (amino acids 126–145, in to out; 423–443, out to in), while AHK2 is predicted to have four transmembrane segments (amino acids 31–47 , in to out; 172–193, out to in; 234–251, in to out; 539–558, out to in) according to the TMpred program: www.ch.embnet.org/software/TMPRED_form.html. We found that expression of part of the cDNA corresponding to amino acids 196–1176 of AHK2 (AHK2Δ), which lacks the two N-terminal transmembrane segments, had essentially no background growth in the absence of cytokinins while giving very sensitive cytokinin-responses (to as low as 1 nM isopentenyladenine) **(Fig. 9.3)**.

Among the 30 or so phosphorelay signal transduction systems in *E. coli*, the RcsC (His-kinase)-YojN (HPt domain protein)-RcsB (transcriptional response regulator) system is unique, in that it resembles the AHKs (His-kinases)-AHPs (HPt domain protein)-ARRs (type-B transcriptional response regulator) cytokinin signal

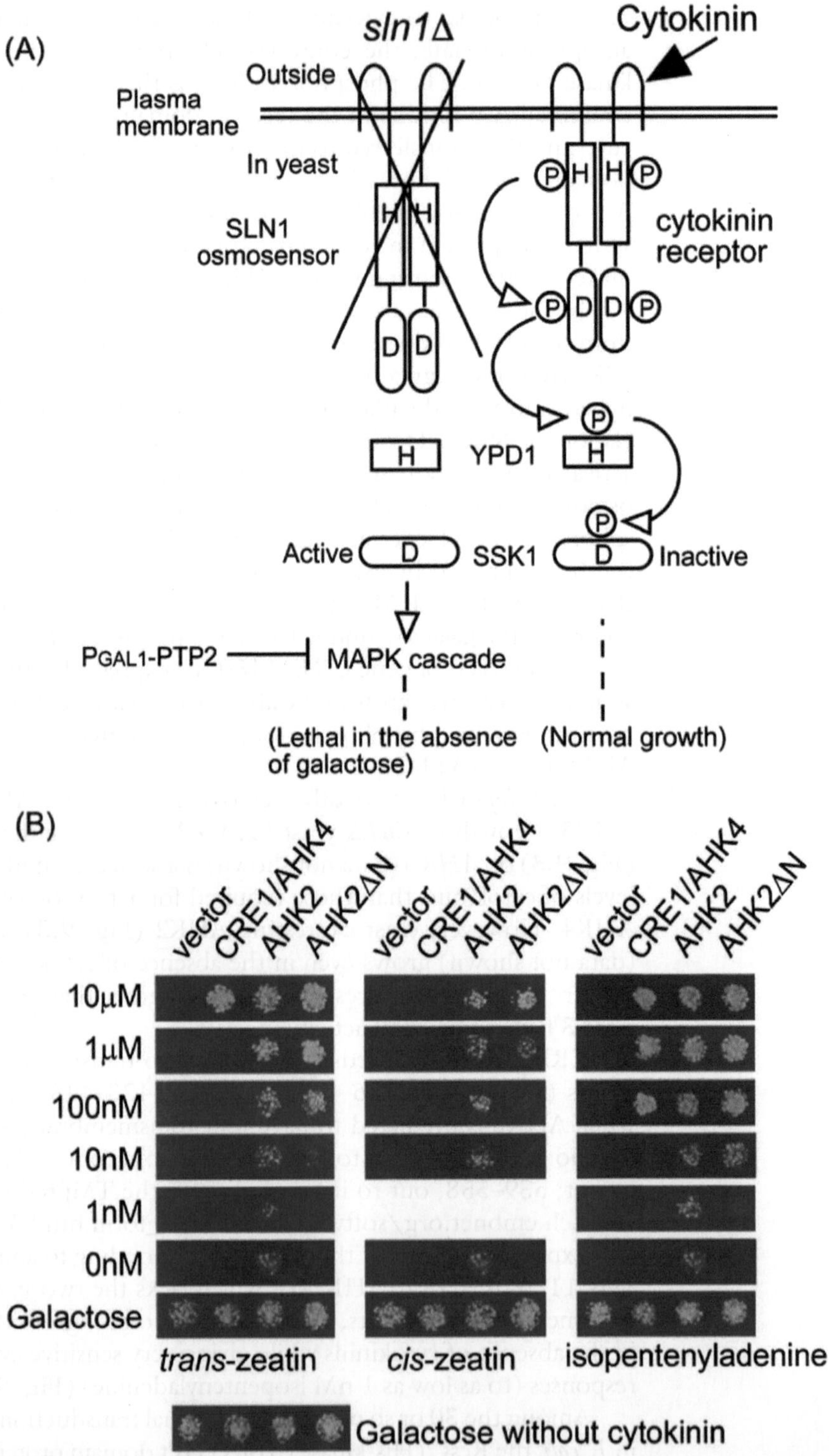

Fig. 9.3. (**A**) The cytokinin-sensing system created in yeast. Cytokinin-receptors suppress the lethality of TM182 (*sln1*Δ) in a cytokinin-dependent manner. TM182 is lethal because the dephosphorylated SSK1 constitutively activates the MAPK

transduction pathway *(14, 15)*. Indeed, it was found that the *CRE1/AHK4* gene has the ability to complement an *E. coli rcsC* deletion mutant; provided that the plant His-kinase protein was properly expressed in *E. coli* cells. To measure this complementation quantitatively, a unique *E. coli* mutant strain (named KMI001 lacking the *rcsC* gene) is used *(14)* in which the *lacZ* reporter gene provided an output for the Rcs-phosphorelay activation. When this *E. coli* mutant is transformed with an appropriate expression plasmid carrying the *CRE1/AHK4* gene (*see* **Fig. 9.2**), the expression of *lacZ* is induced in response to cytokinins through the artificial CRE1/AHK4-mediated Rcs-phosphorelay pathway. Thus, this *E. coli* system provides a way to measure the CRE1/AHK4 cytokinin receptor activity in a highy quantitative manner.

3.1. Cytokinin Response Assay in Yeast

1. Streak TM182 carrying a plasmid with a cytokinin receptor gene (*see* **Note 1**) on -ura/-leu/galactose medium and incubate at 30°C for 36–48 h.
2. Pick a colony and suspend in 1 ml of –ura/-leu/glucose medium and then count the cell density with a haemocytometer.
3. Make a dilution series with –ura/-leu/glucose medium ($1 \times 10^5, 5 \times 10^4, 2.5 \times 10^4$ and 1.25×10^4 cells per milliliter)
4. Spot 10 μl of every dilution on –ura/-leu/glucose with and without cytokinin, and on –ura/-leu/galctose medium.
5. Incubate the yeast plates at 30°C for 36–48 h.
6. The data can be assessed based on a dilution that gives the appropriate number of colonies on galactose-containing medium (*see* **Note 3**).

3.2. Semi-quantitative Measurement of the CRE1/AHK4 Cytokinin Receptor Activity in E. coli

1. The *E. coli* strain KMI001 carrying pINΔEH-AHK4 is cultivated overnight in LB at 37°C.
2. The fully-grown *E. coli* cells are diluted with fresh LB (to approximately 10^7 cells/mL).
3. The diluted *E. coli* cells are spotted onto LB agar-plates containing X-Gal, ampicillin (50 μg/mL), and varying concentrations of *trans*-zeatin (0.01–10 μg/mL).
4. The plates are incubated for 48 h at 25°C.

Fig. 9.3 (continued) pathway. Cytokinins activate the histidine kinase activity of the cytokinin receptors to initiate the phosphorelay, whereby the phosphoryl group is transferred from the activated a cytokinin-receptor to YPD1, then to SSK1, suppressing the lethality of TM182. (**B**) Growth of TM182 carrying CRE1/AHK4, AHK2 or AHK2ΔN in response to different cytokinin concentrations. *trans*-zeatin and isopentenyladenine exhibit potent cytokinin activity, while *cis*-zeatin exhibits weak activity. This order of potency corresponds to the potency of cytokinins for *Arabidopsis*. The yeast strain expressing *AHK2* grows in the absence of cytokinins. Yeast expressing *AHK2ΔN* does not grow in the absence of cytokinins while it can respond to low levels of cytokinins.

5. The CRE1/AHK4 receptor activity can be estimated semi-quantitatively on the basis of the intensity of the resulting blue colony color.
6. For a typical result, see Ref. *(8)*.

3.3. Quantitative Measurement of the AHK4/CRE1 Cytokinin Receptor Activity in E. coli

1. The *E. coli* strain KMI001 carrying pINΔEH-AHK4 is cultivated overnight in LB at 37°C.
2. Stationary phase *E. coli* culture (60 μL) is used to inoculate 1.5 mL of fresh LB. In addition to ampicillin (50 μg/mL), varying concentrations of cytokinins are added to the medium (e.g., *trans*-zeatin, benzyl-adenine, isopentenyl-adenine, thidiazuron, etc.) (0.01–10 μM). Then, *E. coli* cells are cultivated overnight at 25°C by vigorously shaking.
3. Optical density (OD600 nm) of the resulting culture is carefully measured.
4. An aliquot (100 μL) of cells is suspended in 400 μL of Z-buffer in a test tube, and then a drop of toluene (10 μL) is added followed by vigorous mixing.
5. After being incubated for 1 h at 37°C, β-galactosidase activity is measured at 30°C, as follows. To start the reaction, ONPG (100 μL) is added. When the color of reaction mixture turns yellow, 1 M Na_2CO_3 (1 mL) is added to stop the reaction. Record the reaction time as precisely as possible (usually 10–30 min).
6. The optical density of the reaction mixture is measured at both 420 and 550 nm.
7. β-Galactosidase activity (relative units) is calculated by using the following formula: $1000 \times [(\mathrm{OD420}-\mathrm{OD550} \times 1.75)/t\,(\mathrm{min})\ \mathrm{vol.}\ (\mathrm{mL})\ \mathrm{OD600}]$.
8. A typical result is shown in **Fig. 9.2** (see also Ref. *(4)*).

4. Notes

1. To store yeast strains, suspend yeast in sterile 10% glycerol and store at –80°C. Do not repeat freeze and thaw cycles.
2. The TM182 strain grows only in the presence of galactose, which induces the *PTP2* gene. The TM182 strain with or without a cytokinin receptor gene often produces suppressor mutants that grow in the absence of galactose and cytokinin. Frozen stocks should be streaked on gatactose-containing dropout medium and use a single colony. Transform plasmids carrying a cytokinin receptor if necessary (i.e., you suspect that a frozen stock has reverted).
3. Yeasts grow slower on media containing galactose than glucose.

4. An appropriate yeast density is 100–200 CFU (colony forming/10 μl). To determine CFU, 100 μl of the dilution series are spread onto –ura/-leu/galactose plate, colonies are allowed to grow and counted. Divide the colony number by 10.
5. The *E. coli* mutant strain used is versatile in the sense that one can also measure the His-kinase activities of AHK2 and AHK3 in a quantitative manner, provided that an appropriate expression vectors were constructed *(5, 16)*. One can examine any putative cytokinin recptor His-kinase genes from plants other than *A. thaliana (16)*.
6. The yeast and *E. coli* systems are even more versatile in that one can generally examine other *Arabidopsis* His-kinase genes, such as CKI1 and AHK5/CKI2, both of which function in a manner independent of cytokinin *(5, 17)*.

Acknowledgment

We thank Dr. Saito, H. (current affiliation: Department of Basic Medical Sciences, Institute of Medical Science, University of Tokyo) for providing us with TM182.

References

1. Kakimoto, T. (2003) *Annu Rev Plant Biol* **54,** 605–27.
2. Mok, D. W., and Mok, M. C. (2001) *Annu Rev Plant Physiol Plant Mol Biol* **52,** 89–118.
3. Sakakibara, H. (2006) *Annu Rev Plant Biol* **57,** 431–49.
4. Spichal, L., Rakova, N. Y., Riefler, M., Mizuno, T., Romanov, G. A., Strnad, M., and Schmulling, T. (2004) *Plant Cell Physiol* **45,** 1299–305.
5. Yamada, H., Suzuki, T., Terada, K., Takei, K., Ishikawa, K., Miwa, K., Yamashino, T., and Mizuno, T. (2001) *Plant Cell Physiol* **42,** 1017–23.
6. Stock, A. M., Robinson, V. L., and Goudreau, P. N. (2000) *Annu Rev Biochem* **69,** 183–215.
7. Inoue, T., Higuchi, M., Hashimoto, Y., Seki, M., Kobayashi, M., Kato, T., Tabata, S., Shinozaki, K., and Kakimoto, T. (2001) *Nature* **409,** 1060–3.
8. Suzuki, T., Miwa, K., Ishikawa, K., Yamada, H., Aiba, H., and Mizuno, T. (2001) *Plant Cell Physiol* **42,** 107–13.
9. Mahonen, A. P., Higuchi, M., Tormakangas, K., Miyawaki, K., Pischke, M. S., Sussman, M. R., Helariutta, Y., and Kakimoto, T. (2006) *Curr Biol* **16,** 1116–22.
10. Romanov, G. A., Spichal, L., Lomin, S. N., Strnad, M., and Schmulling, T. (2005) *Anal Biochem* **347,** 129–34.
11. Maeda, T., Wurgler-Murphy, S. M., and Saito, H. (1994) *Nature* **369,** 242–5.
12. Mumberg, D., Muller, R., and Funk, M. (1995) *Gene* **156,** 119–22.
13. Posas, F., Wurgler-Murphy, S. M., Maeda, T., Witten, E. A., Thai, T. C., and Saito, H. (1996) *Cell* **86,** 865–75.
14. Takeda, S., Fujisawa, Y., Matsubara, M., Aiba, H., and Mizuno, T. (2001) *Mol Microbiol* **40,** 440–50.
15. Hagiwara, D., Sugiura, M., Oshima, T., Mori, H., Aiba, H., Yamashino, T., and Mizuno, T. (2003) *J Bacteriol* **185,** 5735–46.
16. Tirichine, L., Sandal, N., Madsen, L. H., Radutoiu, S., Albrektsen, A. S., Sato, S., Asamizu, E., Tabata, S., and Stougaard, J. (2007) *Science* **315,** 104–7.
17. Iwama, A., Yamashino, T., Tanaka, Y., Sakakibara, H., Kakimoto, T., Sato, S., Kato, T., Tabata, S., Nagatani, A., and Mizuno, T. (2007) *Plant Cell Physiol* **48,** 375–80.

Chapter 10

Hormone-Binding Assay Using Living Bacteria Expressing Eukaryotic Receptors

Georgy A. Romanov and Sergey N. Lomin

Abstract

Studies on hormone-receptor interaction include, as a rule, isolation and extensive purification of the receptor protein or a particular receptor-containing fraction. To bypass these time- and resource-consuming procedures, we proposed a live cell-based assay using transgenic bacteria expressing single eukaryotic receptors. We describe here ^{3}H-cytokinin binding to corresponding plant receptors as an example. The method includes procedures of bacteria growing, incubation with labeled hormone, separation of bound from unbound ligand, determination of radioactivity in bacterial precipitates, and mathematical analysis of primary data. The established simple protocol for specific labeling hormone-binding sites in intact bacteria allows determination of the main parameters of the ligand–receptor interaction.

Key words: Receptor, hormone, ligand, binding assay, binding parameters, transgenic bacteria.

1. Introduction

The interaction of hormone with a cognate receptor is a first and crucial stage of hormone signaling in the cell. Therefore, the characteristics of such an interaction (binding) are of great importance for the subsequent hormone-triggered intracellular events. In particular, the high affinity to hormone of a presumable sensor protein serves as important argument for its receptor function. The hormone–receptor interaction is usually studied using labeled hormone and isolated receptor protein or a particular receptor-containing fraction *(1)*. In most cases, the procedure of receptor or particular fraction isolation and purification requires time, expensive equipment, and chemicals. The subsequent binding assay with purified receptor or

Sean Cutler, Dario Bonetta (eds.), *Plant Hormones: Methods and Protocols, Second Edition, vol. 495*

DOI 10.1007/978-1-59745-477-3_10 Springerprotocols.com

fraction also requires special equipment to discriminate between bound and unbound hormone. These difficulties become even harder in studies of membrane receptors which often lose both stability and hormone-binding capacity upon purification *(2)*.

Some attempts to avoid stages of receptor purification in studies of hormone perception have already been made. In particular, the recently developed tissue-segment binding method is based on the use of living tissue microsegments (of mm size) in a physiological solution *(3)*. In this way, receptors remain in conditions approximating natural ones and the binding assay is significantly simplified as preliminary tissue fractionation is no longer needed. The shortcomings of this method are associated with processes that might exert an influence on the labeled hormones, receptors, or the interaction between them. These processes include metabolism and intracellular transport of the labeled ligand, proteolysis and/or internalization of receptors and/or ligand-receptor complexes. The competition of endogenous hormone with the labeled ligand for the receptor is also a concern. In addition, natural tissue segments may contain different receptor moieties for the same hormone, making it difficult to discriminate among different receptor types.

We have used intact transgenic bacteria (*E. coli*) expressing genes encoding functionally active eukaryotic receptors. Sensor histidine kinases CRE1/AHK4 and AHK3 from Arabidopsis represent examples of such receptors. We took advantage of the fact that these receptors were shown to enable *E. coli* to sense cytokinin *(4–6)* and hence retain their functionality. Here we describe the use of living transgenic bacteria for direct binding studies, i.e. to characterize the interaction of the receptor with cognate ligands. The method includes procedures for growing bacteria, their incubation with highly labeled hormone, separation of bound from unbound hormone, determination of radioactivity in bacterial precipitates, and mathematical analysis of primary data. It is important to note that our binding assays were done under conditions (low temperature, short time) preventing metabolic degradation of ligands or receptors. Therefore, the results reflect genuine hormone–receptor interactions and are not confounded by any kind of cellular metabolism. Our data *(7–9)* show that this rapid and inexpensive approach allows the determination of the most important characteristics of the hormone–receptor interaction.

2. Materials

2.1. Bacterial Strains and Chemicals for Culturing

1. *Escherichia coli* KMI001 strain carrying either one of two vectors, pINIIIΔEH expressing *CRE1/AHK4 (4, 5, 7)* or pSTV28 expressing *AHK3 (6, 8)*. KMI001 carrying the

empty vector pINIIIΔEH is used as a control (*see* **Note 1**). Also, *E. coli* BL21 strain (Life Technologies) carrying the receptor-expressing vector pDEST-AHK (Invitrogen) *(9)* can be used.

2. Standard M9 medium (pH 7) supplemented with 0.1% casamino acids (Difco, USA) *(10)* (*see* **Note 2**).
3. Bacteria are kept as a frozen stock at –70°C or for a few weeks as colonies on agar-solidified LB medium supplemented with the appropriate antibiotic at 2–4°C.
4. Stock solutions of the antibiotics carbenicillin (50 mg/ml in water) and chloramphenicol (25 mg/ml in ethanol) should be kept at –20°C *(10)*.

2.2. Materials for Binding Assay with Labeled Hormone

1. *Trans*-[2-^{3}H]zeatin preparations with high specific activity (600–900 GBq/mmol) should be obtained (*see* **Note 3**). We have obtained these from the Isotope laboratory of the Institute of Experimental Botany (Prague, Czech Republic); other commercial sources of labeled zeatin also exist (e.g., American Radiolabeled Chemicals Inc.).
2. For experiments, 50-fold dilutions of ^{3}H-zeatin are made in 20% ethanol and can be stored at –15°C in the darkness for at least 1–2 months. 2.5 μl of this dilution contains about 4×10^4 cpm of radioactivity and corresponds to approximately 2 pmole of hormone.
3. Cytokinins and related compounds (OlChemim, Olomouc, Czech Republic and/or Sigma Aldrich) should be dissolved in DMSO as stock solutions (conc. 0.1 M) (*see* **Note 4**) and stored until use at –20°C (*see* **Note 5**).
4. Dilutions of cytokinins and other ligands should be prepared in advance. For most ligands, 40–50% DMSO in water is very suitable as a solvent (*see* **Note 6**).
5. Precise micropipettes for μl volumes.

2.3. Separation of Bound and Unbound Hormones

1. A device for vacuum aspiration of supernatant from eppendorf tubes, with a trap for radioactive solutions (like a stoppered side flask).
2. Ice-bath and a tube shaker for a set of eppendorf tubes.
3. High quality ethanol (we use pure 99% ethanol).
4. High-throughput dispenser for ethanol.

2.4. Determination of Bound Radioactivity

1. Scintillation cocktail in a suitable dispenser (5–20 ml) (*see* **Note 7**).
2. Scintillation counter and other corresponding equipment (vials, racks, etc.).

2.5. Analysis of Primary Data

1. Excel or similar computer programs for primary data storage and analysis.
2. SigmaPlot or other specialized computer programs for binding parameter determination.

3. Methods

One of the most important tasks in a study of hormone–receptor complex formation is the quantitative discrimination between bound and unbound hormone. In turn, specific and non-specific binding also needs to be differentiated *(1, 2)*. The common approach is to make the receptor insoluble while the free hormone remains soluble. Soluble and insoluble phases are easily separated, for example, by centrifugation, allowing the hormone fraction bound to receptors and the unbound one to be quantitatively discriminated. When a receptor is expressed inside bacteria it remains in the insoluble phase while unbound hormone is present mainly in the soluble phase, except for a minor part which can be mechanically trapped by bacterial cells (the so-called non-specific binding). Receptors which are expressed in bacteria are easily accessible to hormones of low mol.wt (below 600–900 kD) and the bacterial suspension can be quantitatively aliquoted to produce standard probes. These properties of receptor-synthesizing bacteria provide all the necessary prerequisites for a simple binding assay *(7)*.

3.1. Preparation of Bacteria

1. Under sterile conditions, inoculate 1–4 ml of freshly prepared M9, supplemented with carbenicillin or chloramphenicol (in our case, 25 μg/ml of antibiotic) with a single colony of transformed bacteria. Take 5–10 individual clones from a Petri dish. Make 2–3 replicates for each clone. Incubate tubes inclined overnight at 22–24°C in darkness with vigorous shaking (*see* **Note 8**).
2. Let bacteria grow until optical density at 600 nm (OD_{600}) of the overnight pre-culture reaches approximately 1. Dilute more dense cell suspensions with M9.
3. Check the ^{3}H-hormone binding of each clone according to **Sections 3.2–3.5**. Select clones of maximal specific binding capacity for the subsequent step (*see* **Note 8**).
4. Inoculate a large volume of M9 containing the appropriate antibiotic with 50–100 μl of the selected pre-culture(s). The volume for 24 probes corresponds to approximately 25–30 ml; for 48 probes about 50–60 ml.
5. Incubate the bacteria overnight at 22–24°C in darkness with vigorous shaking. Take small aliquots to read OD_{600}. The optimal final concentration of the suspension corresponds to an OD_{600} of about 1–1.2.

6. Keep the bacteria before experiment in a refrigerator (avoid freezing!). For all subsequent manipulations, the culture should be kept at a low temperature (ice bath).

3.2. Incubation of Bacteria with Hormone(s) (See Note 9)

1. Put 2.5 μl of 50-fold diluted ^{3}H-zeatin (around 2 pmole) in a series of standard eppendorf tubes. Aliquot these as precisely as possible. Protect the tubes from intense light (*see* **Note 10**).
2. To test non-specific binding or radioactivity displacement, add non-labeled hormone(s) together with the ^{3}H-zeatin (1.25 μl of 0.16–16 mM solution). To probes containing only ^{3}H-zeatin, add 1.25 μl of the non-labeled diluent (DMSO) (*see* **Note 11**).
3. Put all the eppendorfs with ^{3}H-zeatin in an ice bath. Protect the tubes from intense light.
4. Transfer the bacterial culture from the refrigerator to an ice bath. Swirl the culture manually for few seconds to obtain a homogenous suspension.
5. Add precisely 0.75 ml of the homogenous suspension in each eppendorf containing labeled hormone (*see* **Note 12**). When pipetting bacteria, swirl the suspension periodically to ensure that the bacteria remain suspended. Close the lids tightly. Again, take care to protect the tubes from intense light.
6. Mix each suspension gently by inverting the tubes three times.
7. Put the tubes on ice (in the dark) for 20–30 min. Repeat the mixing (1 inversion) after every 10 min of incubation.

3.3. Separation of Bound from Unbound Hormone

1. Following the incubation step, centrifuge the tubes for 2–3 min at maximum rpm (usually about 13000 rpm or 16000 × g) in a pre-cooled centrifuge (or pre-cooled rotor).
2. Transfer the tubes carefully from the centrifuge to an ice bath.
3. Remove all of the supernatant by vacuum aspiration. Do not touch the pellets! (*see* **Note 13**).
4. Put the tubes containing the bacterial pellets in a rack and keep them at room temperature (RT). Fill the tubes with 0.24 ml of high quality ethanol (*see* **Note 14**).
5. Close the tubes tightly immediately after adding the ethanol and shake them vigorously by vortexing (1400 rpm in an Eppendorf thermomixer, Germany) at RT for at least 1 h.
6. (Optional) Brief centrifugation (~ 1 min) at 13000 rpm, RT (*see* **Note 15**).

3.4. Determination of Bound Radioactivity

1. Transfer the eppendorfs to a rack at RT.
2. Transfer exactly 0.2 ml of each supernatant to scintillation vials (*see* **Notes 14, 16**).
3. Close the vials to minimize ethanol evaporation.

4. Fill the vials with 5 ml of ACS-II scintillation cocktail (Amersham Biosciences, UK) (*see* **Note 17**).
5. Put the vials in a scintillation counter and count each vial for 5–10 min. Make sure that the device is programmed for ^{3}H counting (*see* **Note 18**).

3.5. Analysis of Primary Data

1. Copy all the numeric data from the counter to an Excel table (*see* **Note 19**).
2. Calculate the total and the non-specific binding. Verify the significance of the difference between the total and non-specific binding (*see* **Note 20**).

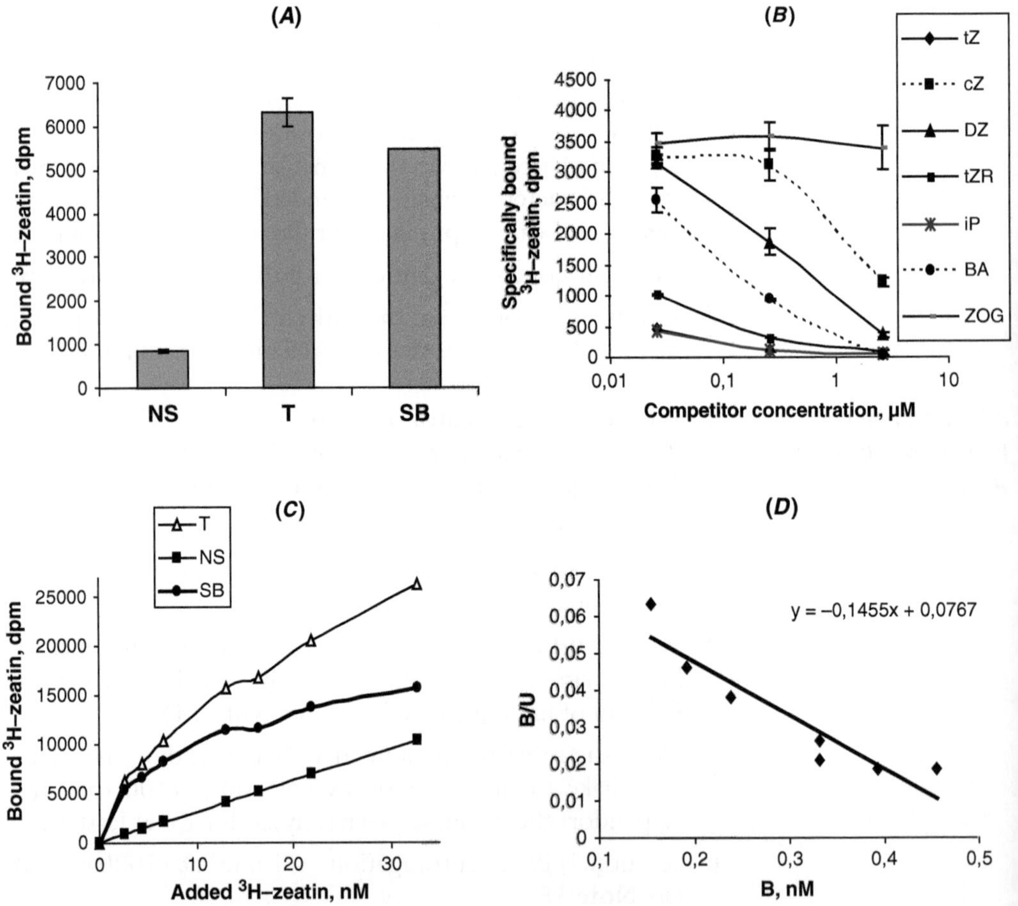

Fig. 10.1. Some examples of binding assays with *E.coli* clone BL21 pDEST (Invitrogen), expressing the cytokinin receptor CRE1/AHK4. Bacteria were grown overnight in M9 medium. For standard assays, ^{3}H-zeatin (851 GBq/mmol, about 2 pmole in 2.5 μl) was used. All incubations were on ice for 30 min. (**A**) Non-specific (NS), total (T) and specific (SB) binding. (**B**) Displacement of ^{3}H-zeatin with different unlabeled cytokinins and derivatives (ligand specificity). (**C**) Dose-dependence of total (T), non-specific (NS) and specific (SB) binding. (**D**) Scatchard plot produced with the data from (C); B indicates specifically bound-, and U, unbound hormone.

3. When the specific binding is significant, is reproducible and high, other calculations can be made depending on the aim of the experiment (*see* **Note 21**) (typical examples of results are shown in **Fig. 10.1.**

4. Notes

1. It should be taken into account that not all bacteria expressing eukaryotic receptor gene are suitable for the direct binding assay. The prerequisite for the assay is that the receptor is folded correctly in the bacterial cell and does not require posttranslational modification or additional protein(s) for its binding activity. Also, the degree of receptor expression in the bacteria might be critical.
2. M9 is a convenient medium composed only of mineral salts. In our experience, large modifications of the medium are possible; M9 could be replaced by LB medium, for example. However, one should be cautious regarding the pH of the medium as some plant receptors might have an optimal pH that is not in the neutral range. Changing the pH can be easily achieved after bacterial precipitation and washing with the corresponding buffer solution.
3. This labeled compound is delivered in ethanol and must be stored in a freezer (–20°C) in darkness. Nevertheless, even when stored correctly the ^{3}H-labeled ligands slowly decompose mainly due to radiolysis. For qualitative assays, ^{3}H-ligands stored for over several months can be used, but for quantitative assays freshly synthesized ^{3}H-ligands (stored less than one month) are preferable. In general ligands (hormones) used for the binding assay should be highly radioactive (specific radioactivity not less than 500–600 GBq/mmol), stable, homogenous and possess high affinity and selectivity toward the receptor being studied (see *(11)* for more information).
4. Some compounds that are less soluble in DMSO can be dissolved at lower concentration (10 mM) as well. To assess non-specific binding, one usually uses a 100- to 10000-fold excess of non-labeled ligand over the labeled one.
5. Attention should be paid to possible precipitation of concentrated hormones during their storage. If a precipitate appears, the stock solution should be gently warmed before use to ensure complete dissolution of the compound (*see* also **Note 6**).
6. The mixture of DMSO and water has the advantage that it remains liquid at rather low temperatures. Unlike pure water or pure DMSO, solutions of 40–60% DMSO remain liquid even at –20°C. This precludes the necessity of thawing solutions

every time before their use. However, always check for the absence of precipitate in probes with concentrated ligand! (*see* also **Note 5**).

7. Not all dispenser types are suitable for scintillation liquids which usually contain strong solvents like dioxane or toluene. We use a solvent-resistant dispenser; Ceramus 2–10 ml from Hirschmann Labogerate (Germany).
8. Not all individual clones from the same culture display similar binding activity. In some clones, the receptor might become mutated or weakly expressed. Select clones with high and stable activity for subsequent experiments. Store these clones as a frozen stock at –70°C for the future work.
9. Work with radioactivity requires strict precautions. It is necessary to wear gloves and work in a special room with separate and marked equipment. All tritium-contaminated waste must be collected separately in a special container. Follow the regulations in your jurisdiction for radioactive waste handling and disposal.
10. Usually a set of 24 or 48 eppendorfs are used in each one of our experiments. We recommend marking the tubes with numbers (1,…, 24 or 1,…, 48) and running all the manipulations in the same order, i.e., from the tube 1 up to the last one; this ensures an equal temporal processing of probes. Light protection is necessary for zeatin, since it is sensitive to light.
11. It is desirable to add hormones in small volumes because this minimizes the DMSO concentration in the final mixture. DMSO concentrations higher than 1% could significantly influence the hormone–receptor interaction. For this reason, the final DMSO concentration should ideally be kept to less than 0.1%.
12. Seven-hundred and fifty microliters is a convenient volume for 1.5-ml eppendorfs, but other volumes might be suitable as well depending on the bacterial concentration and their binding properties. We have successfully tested ligand binding by use of bacterial suspensions in the volume range of 0.5–1.2 ml. Do all manipulations with transgenic bacteria wearing gloves in a hood (sterile conditions are not necessary). Collect used transgenic cultures and used tips in a special container for subsequent sterilization and disposal.
13. Following aspiration of the supernatant, some drops might still remain on the tube walls. Try to remove them as much as possible. Sometimes it is better to do the aspiration in two stages; the first one to remove the bulk of the supernatant and then the second one, which can be done following pulse centrifugation, to remove the remaining traces of liquid.

14. The use of ethanol is advantageous over SDS (1%) because the latter dissolves all cellular material, which ends up in the scintillation vials reducing the efficiency and stability of counting. It is also possible to extract the radioactivity with other amounts of ethanol. For example, 0.2 ml can be used and the entire volume can be transferred to scintillation vials for counting. This latter procedure may however take longer time.
15. If the pellets remain solid after extraction this operation can be omitted.
16. Different vial types can be used. Mini-vials of 5 ml capacity are quite convenient, but standard vials (of 20 ml) are also suitable. In the latter case, 5–10 ml of scintillation fluid is sufficient for counting radioactivity dissolved in a small volume of ethanol.
17. Other scintillation fluids can be equally employed. In our laboratory, we have used ZhS-8 (ЖС-8) (Reahim, Russia), a dioxane-type scintillator.
18. In practice, we often employ two counting programs; a quick initial count for 1 min to assess the quality of experiment and then, if necessary, a longer count (5–10 min) to obtain more precise data. The 1–2 h incubation (RT) of vials before counting upon mixing probes with scintillation liquid allows ^{3}H-zeatin to redistribute homogenously in the vials thus improving the accuracy of counting. Hermetically closed vials can be kept at RT for weeks and recounted again if necessary.
19. Other computer programs designed for data storage and transformation can be equally employed.
20. The specific binding is only possible in cases when a significant difference exists between the total and non-specific binding. The greater the difference, the more reliable and quantitative the binding characteristics. Optimization of the binding assay should reduce non-specific binding as much as possible, leaving the total binding at a level of 1000–10000 dpm. For a reliable quantitative analysis, the non-specific binding should never exceed 20–30% of the total one (see *(11)*).
21. For binding parameter assessment, we use Excel and/or SigmaPlot8; other software is also suitable for these calculations.

Acknowledgments

We would very like to thank T. Schmülling (Germany) for providing all favorable conditions to accomplish some of the experiments in his laboratory; T. Mizuno, H. Sakakibara (Japan), A. Heyl and K. Wulfetange (Germany) for their kind gift of bacterial clones expressing cytokinin receptors. Our work is supported by grants RFBR (Russia) NN 04-04-49120, 07-04-00331 и 07-04-91211-ЯФ, and DAAD (Germany) for S.L.

References

1. Gomperts, B.D., Kramer, I.M., and Tatham, P.E.R. (2003) *Signal Transduction*. Academic Press, San Diego et al.
2. Cuatrecasas, P., and Jacobs, S. (eds.) (1981) *Membrane Receptors, Methods for Purification and Characterization*. Chapman and Hall, London, N.Y.
3. Muramatsu, I., Tanaka, T., Suzuki, F., Li, Z., Hiraizumi-Hiraoka, Y., Anisuzzaman, A.S.M., Yamamoto, H., Horinouchi, T., and Morishima, S. (2005) Quantifying receptor properties: the tissue segment binding method – a powerful tool for the pharmacome analysis of native receptors. *J. Pharmacol.* **98**, 331–339.
4. Suzuki, T., Miwa, K., Ishikawa, K., Yamada, H., Aiba, H., and Mizuno, T. (2001) The *Arabidopsis* sensor His-kinase, AHK4, can respond to cytokinins. *Plant Cell Physiol.* **42**, 107–113.
5. Yamada, H., Suzuki, T., Terada, K., Takei, K., Ishikawa, K., Miwa, K., Yamashino, T., and Mizuno, T. (2001) The *Arabidopsis* AHK4 histidine kinase is a cytokinin-binding receptor that transduces cytokinin signals across the membrane. *Plant Cell Physiol.* **42**, 1017–1023.
6. Spíchal, L., Rakova, N.Y., Riefler, M., Mizuno, T., Romanov, G.A., Strnad, M., and Schmülling, T. (2004) Two cytokinin receptors of *Arabidopsis thaliana*, CRE1/AHK4 and AHK3, differ in their ligand specificity in a bacterial assay. *Plant Cell Physiol.* **45**, 1299–1305.
7. Romanov, G.A., Spíchal, L., Lomin, S.N., Strnad, M., and Schmülling, T. (2005) A live cell hormone-binding assay on transgenic bacteria expressing a eukaryotic receptor protein. *Analyt. Biochem.* **347**, 129–134.
8. Romanov, G.A., Lomin, S.N., and Schmülling, T. (2006) Biochemical characteristics and ligand-binding properties of Arabidopsis cytokinin receptor AHK3 compared to CRE1/AHK4 as revealed by a direct binding assay. *J. Exp. Botany*, **57**, 4051–4058.
9. Heyl, A., Wulfetange, K., Pils, B., Nielsen, N., Romanov, G.A., and Schmülling, T. (2007) Evolutionary proteomics identifies amino acids essential for ligand-binding of the cytokinin receptor CHASE domain. *BMC Evolutionary Biol.* **7**, 62
10. Sambrook, J., and Russell, D.W. (2001) *Molecular Cloning: a Laboratory Manual*, 3rd edn. Cold Spring Harbor Laboratory Press, Cold Spring Harbor, N.Y.
11. Lomin S.N., and Romanov, G.A. (2008) The analysis of hormone-receptor interaction. Theoretical and practical aspects. *Russian J. Plant Physiol.* **55**, 259–273

Chapter 11

Discovery of Plant MicroRNAs and Short-Interfering RNAs by Deep Parallel Sequencing

Padmanabhan Chellappan and Hailing Jin

Abstract

Endogenous small RNAs, including microRNAs (miRNAs) and small interfering RNAs (siRNAs), play important regulatory roles in development, hormone signaling, stress responses, and genome stability. These small RNAs induce transcriptional or posttranscriptional gene silencing by guiding heterochromatin formation, mRNA degradation, or translational inhibition. In this chapter, we describe the methods for small RNA discovery in plants by small RNA cloning and deep parallel sequencing. We compare two methods of small RNA library construction: the 5' phosphate (P)-independent and 5' phosphate (P)-dependent methods. Deep parallel sequencing of small RNA libraries is discussed by comparing among 454, SBS, and SOLiD technologies.

Key words: miRNA, siRNA, small RNA library construction, 454, SBS.

1. Introduction

Small-interfering RNAs (siRNAs) *(1)* and microRNAs (miRNAs) *(2)* are important regulators of gene expression in both plants and animals *(3–5)*. miRNAs are mainly generated from single-stranded fold-back transcripts, whereas siRNAs arise from perfectly paired, long, and double-stranded RNAs *(6)*. Several subclasses of endogenous siRNAs include chromatin-associated siRNAs, trans-acting siRNAs (tasiRNAs) and natural antisense transcript-generated siRNAs (nat-siRNAs) *(3, 7–10)*. In *Arabidopsis*, small RNA biosynthetic pathways involve four DICER-like (DCL) proteins, six RNA-dependent RNA polymerases (RDRs) and ten ARGONAUTEs (AGOs) *(5)*. DCL proteins are RNase III-like

Sean Cutler, Dario Bonetta (eds.), *Plant Hormones: Methods and Protocols, Second Edition, vol. 495*

DOI 10.1007/978-1-59745-477-3_11 Springerprotocols.com

nucleases and produce small RNAs with distinct 5'-phosphate (P) group and 3'-hydroxyl (OH) group.

Small RNA-directed gene silencing regulates many cellular processes, including hormone signaling, development, stress responses, and genome maintenance. Cloning and deep parallel sequencing of small RNAs from *Arabidopsis thaliana* has revealed hundreds of miRNAs and millions of endogenous siRNAs *(11–18)*. However,

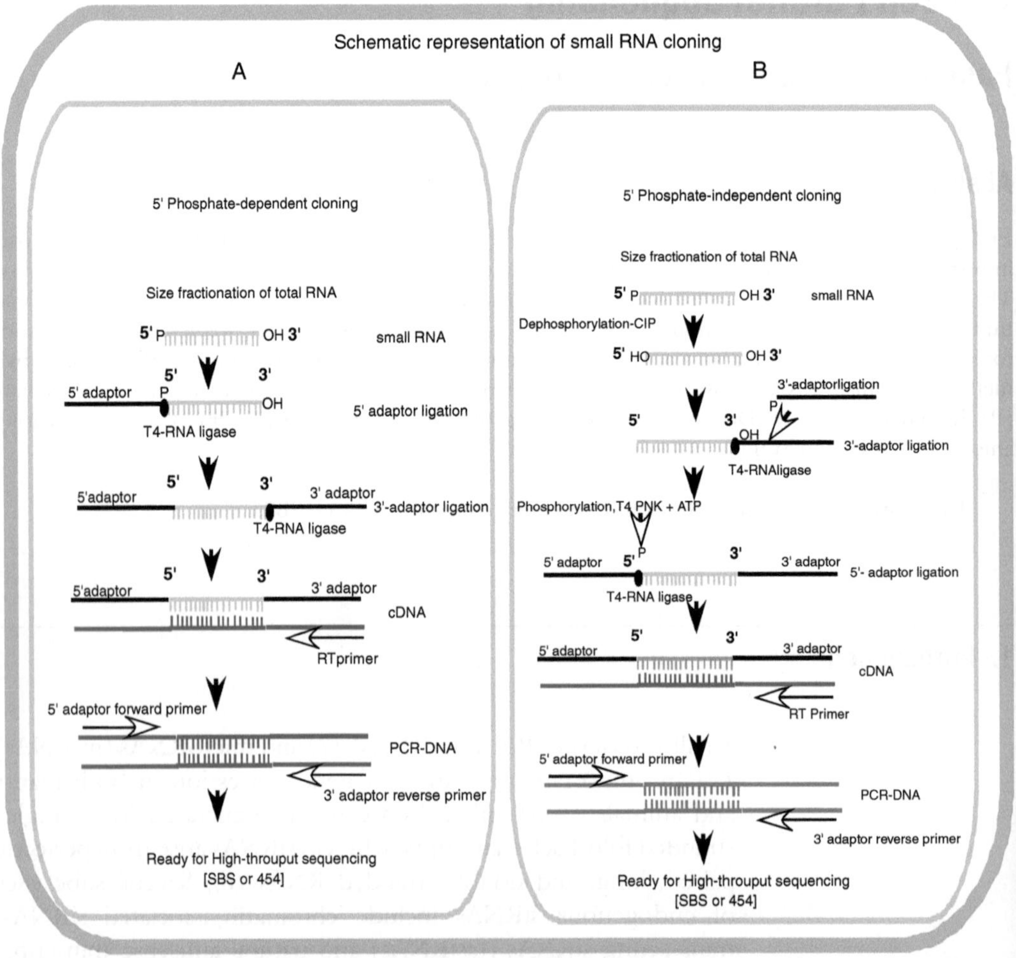

Fig. 11.1. Small RNA cloning: 5' P-dependent and 5' P-independent methods: **(A)** 5' P-dependent method: Small RNAs (20–30 nt) were purified by polyacrylamide gel-based size fractionation. 5'-Adaptor was ligated to the small RNA first, then 3' adaptors was added. Both steps were accomplished by the use of T4-RNA ligase. The cDNA was synthesized by reverse transcription with a primer complementary to 3'-adaptor sequence. PCR was performed to obtain enough DNA for high-throughput deep sequencing. To ensure the accuracy of expression profiling pattern, PCR cycles should be controlled to minimum. The RNA adaptors are indicated in black line, the small RNA in light grey, the cDNA (adaptors + small RNA), and PCR products indicated in dark grey. T4 RNA ligase is indicated by black dot. Arrow indicates location of primers used in PCR. (**B**) 5'P-independent method: 5' dephoshporylation (CIP) was performed to remove the 5'-P group before 3' adaptor ligation to avoid small RNA self-ligation. After 3' adaptor ligation, 5'-P group was added back by phosphorylation using T4 PNK, then 5'-adaptor was ligated. All the subsequent steps are the same as in the 5'P-dependent method (**A**).

there are relatively small overlaps within the datasets generated from different studies, which indicates that discovery of small RNAs is still far away from saturation. The discovery of more small RNAs, especially those induced in response to specific biotic and abiotic stresses or under specific developmental stages, will greatly benefit the understanding of gene regulation in various cellular processes. Here we describe the protocols for constructing small RNA libraries by use of 5' P-independent and 5' P-dependent methods (**Fig. 11.1**) and compare recently developed deep-parallel-sequencing technologies.

2. Materials

2.1. Small RNA Isolation

1. Harvest desired tissue source for RNA isolation using liquid nitrogen and forceps.
2. RNA extraction: Trizol reagent (Invitrogen; 15596-018), chloroform (Fisher Scientific; BP1145-1), isopropanol (Fisher Scientific; BP1145-1), 75% ethanol stored at room temperature, 0.1% diethyl pyrocarbonate (DEPC) (AMRESCO; E174-25G)-treated water stored at room temperature (RT) (*see* **Notes 1** and **2**).
3. Separation of low-molecular-weight (LMW) and high-molecular-weight (HMW) RNA: 5 M NaCl (Fisher Scientific; BP358-212), 50% PEG8000 (Fisher Scientific; BP233-1), 5 mg/ml glycogen (ROCHE; 10901393001).
4. Small-RNA purification: 10X formamide loading buffer (100% formamide, xylene cyanol, and bromophenol blue), 10 bp DNA ladder (1 μg/μl) (Invitrogen; 10821-015), 10% ammonium persulfate (EMD 2310) stored at 4°C, TEMED (Promega; V3161), 40% acrylamide stock (Fisher Scientific; BP1408-1), 0.3 M NaCl, ethanol, ethidium bromide (EtBr).

2.2. Cloning of Small RNAs

5. Dephosphorylation: calf intestinal alkaline phosphatase (CIP) (10 U/μl NEB; M0290S).
6. Phosphorylation: T4 polynucleotide kinase (PNK 10 U/μl, NEB; M0201S).
7. Adaptor ligation: T4 RNA ligase (5 U/μl) and 10X RNA ligase buffer (Ambion; 2140), RNaseOUT (40 U/μl, Invitrogen; 10777).
8. RT-PCR: Superscript II RT (200 U/μl, Invitrogen; 18064), Phusion high-fidelity DNA polymerase (2 U/μl, NEB; F-530S), dNTP mix (Invitrogen; R7525-01).

9. Cloning: TOPO-TA cloning vector and TOP10 one-shot competent cells (Invitrogen; K4500-01), LB Broth Base (Invitrogen; 12780), X-gal (Invitrogen; 15520-034).

3. Methods

3.1. Extraction and Purification of Small RNAs from Plant Tissue

(Suitable for Arabidopsis, Nicotiana, Potato, Tomato, Cassava, etc.)

1. Harvest tissue in a tube containing liquid nitrogen (*see* **Note 3**).
2. Grind tissue to a fine powder in liquid nitrogen (N) using a mortar and pestle (*see* **Note 4**).
3. For 1 gram (g) tissue, add 10 ml Trizol reagent and 2 ml chloroform directly into the mortar and mix well by grinding (in a chemical hood) (*see* **Note 5**).
4. Transfer the liquid into a fresh tube, mix well by vortexing and leave at RT for about 10 min.
5. Centrifuge at 8000–10,000 rpm for 20 min and transfer aqueous phase into a fresh tube (*see* **Note 6**).
6. Extract with an equal volume (Vol.) of phenol:chloroform:-isoamyl alcohol (25:24:1), then precipitate RNA by adding 1/10th Vol. of 3 M sodium acetate and 2.5 Vol. of 100% ethanol or 1 Vol. of 100% isopropanol.
7. Centrifuge samples at maximum speed (>10,000 rpm) in a refrigerated centrifuge for 20 min to pellet out the precipitated RNA, and wash the pellet with 80% ethanol. After a brief air drying, dissolve the RNA in 50% formamide (for gel purification only, not for enzymatic reaction) or in autoclaved DEPC-treated water (*see* **Notes 7, 8** and **9**).
8. HMW RNA is precipitated by adding both 50% PEG (MW=8000) to a final concentration of 5% and 5 M NaCl to a final concentration of 0.5 M.
9. Centrifuge the contents for 10 min, then transfer the supernatant and add 2.5 Vol. of ethanol to precipitate LMW RNA. After centrifugation and wash with 70% ethanol, resuspend the RNA in 50% formamide or DEPC-treated H_2O and store at –80°C.

3.2. Gel Fractionation of 20- to 30-nt Small RNAs

1. Measure the concentration and aliquot samples into fresh tubes (we use total RNA at 200–300 μg per sample for a small-RNA library. The amount of RNA used depends on the abundance of the small RNAs in the specific tissue or sample you use).
2. Incubate samples at 65°C for 10 min and snap cool on ice.
3. After a brief spin, add 3–4 μl of 6X bromophenol blue loading dye.

4. Samples are ready to load onto the gel.
5. Make a 15% denaturing polyacrylamide gel (for a 16 cm × 20 cm gel, use 2.5 ml 10X TBE, 18.75 ml 40% acrylamide-bisacrylamide solution (29:1), 21 g urea, 350 μl 10% APS and 17.5 μl TEMED) and incubate at 37°C to dissolve urea and adjust volume to 50 ml. Remove the comb and assemble the gel in a vertical gel tank and fill the upper and lower tanks with enough 0.5X TBE.
6. Pre-run gels for about 30 min.
7. Before loading samples, rinse the wells individually with running buffer to remove urea.
8. Load the samples into the wells and electrophorese at 200–300 V for 2–3 h, remember to include RNA or DNA ladders for size determination stop when the bromophenol blue dye reaches the end of the gel.
9. Remove one of the glass plates and carefully slide the gel into a tray containing EtBr solution for 5 min for staining.
10. Find bands corresponding to expected size range (20–30 nt) in comparison to a reference marker run parallel in the same gel; excise gel as a strip and crush it into small pieces.
11. Add 1 to 2 Vol. of RNA elution solution (0.3 M NaCl) and 2 μl of RNaseOUT (Invitrogen; 10777019); the solution Vol. should cover the gel pieces.
12. Leave the tube under gentle shaking at 4°C overnight.
13. Remove gel slices by decanting liquid into a fresh tube and extract once with phenol:chloroform:isoamyl alcohol (25:24:1).
14. Precipitate RNA from the aqueous phase in 1/10th Vol. NaOAc, 2.5 Vol. ethanol and glycogen at –20°C for more than 1 h. Centrifuge to pellet the RNA and wash with 80% ethanol; after 5–10 min air dry in a laminar-flow hood and re-suspend the RNA in 20 μl DEPC-treated H_2O.

3.3. Small-RNA Cloning

Here we present two commonly used methods of small RNA library construction: 5' P-independent and 5' P-dependent cloning methods (**Fig. 11.1**). The 5' P-independent method has high ligation efficiency and has the potential to identify Dicer-independent novel small-RNA species such as RDRP-generated secondary siRNAs in *C. elegans (19)*. However, this method cannot discriminate degradation products, which can also be cloned. The 5' P-dependent method can discriminate many degradation products and can clone only the small RNAs with a 5' monophosphate, which is the identity for Dicer products. This method is fast but cannot efficiently avoid self-ligation of small RNAs unless a specific mutated version of RNA ligase is used *(20)*. The following describes the step-wise protocols for each method.

3.4. 5'-P-independent Cloning: Dephosphorylation

Dephosphorylation of small RNA is carried out in a 30 μl reaction containing 20 μl purified small RNA, 3 μl 10X buffer, 2 μl CIP (10 U/μl), 4 μl H_2O, and 1 μl RNaseOUT. Incubate the reaction at 37°C for 60 min.

1. Extract once with phenol:chloroform:isoamyl alcohol (25:24:1).
2. Precipitate RNA from aqueous phase in 1/10th Vol. NaOAc and 2.5 Vol. ethanol in the presence of glycogen at −20°C for more than 1 h.
3. Centrifuge to pellet the RNA and wash in 80% ethanol.
4. After 5 min air dry in a laminar hood. Resuspend the pellet in 20 μl DEPC-treated H_2O.

3.5. 3' Adaptor ligation

3' adaptor ligation of small RNA is performed in a 30 μl reaction containing 20 μl purified small RNA, 3 μl 10X buffer, 3 μl T4 RNA ligase (40 U/μl), 2 μl adaptor (25 μM), 1 μl H_2O and 1 μl RNaseOUT. Mix contents by gentle tapping, and after a brief spin, incubate the tube at RT for 6 h. To avoid self-ligation, the 3' adaptor is designed to have an inverted 3' nucleotide, such as idT.

3.6. Purification of Small RNA with 3' Adaptor

1. Run the ligation mixture on a 15% denaturing polyacrylamide gel at 200–300 V for 2–3 h.
2. Stain the gel in EtBr for 5–10 min and view on a UV transilluminator.
3. Excise a gel slice encompassing the size of RNA (3' adaptor + 20–30 nt small RNA) and crush into pieces.
4. Elute RNA from the gel pieces by immersion in 0.3 M NaCl at 4°C overnight.

Precipitate the RNA in the aqueous phase with 1/10th Vol. NaOAc, 2.5 Vol. ethanol and glycogen at –20°C for a couple of hours. Centrifuge to pellet the RNA and wash in 80% ethanol, after air drying dissolve the RNA pellet in 20 μl DEPC-treated H_2O.

3.7. 5'-Phosphorylation

Phosphorylation of 3' adaptor-ligated small RNA is carried out in a 50 μl reaction containing 20 μl purified small RNA, 5 μl 10X PNK buffer, 3 μl T4 polynucleotide kinase (10 U/μl), 10 μl ATP (cold, 5 nmol/μl) 11 μl H_2O, and 1 μl RNaseOUT. Incubate the reaction at 37°C for 30 min. Inactivate the enzyme by incubating the reaction at 65°C for 20 min.

1. Extract once with phenol:chloroform:isoamyl alcohol (25:24:1).
2. Precipitate RNA from the aqueous phase in 1/10th Vol. NaOAc, 2.5 Vol. ethanol and glycogen at –20°C for more than 1 h.
3. Centrifuge at 4°C to pellet the RNA and wash in 80% ethanol; after 5 min air dry under a laminar hood, resuspend the pellet in 20 μl DEPC-treated H_2O.

3.8. 5'-Adaptor ligation

Adaptor ligation of small RNA is carried out in a 30-μl reaction containing 20 μl purified small RNA, 3 μl 10X buffer, 2 μl T4 RNA ligase (40 U/μl), 2 μl adaptor (25 μM), 2 μl H_2O and 1 μl RNaseOUT; mix gently and incubate at RT for 6 h.

1 Run the ligation mixture on a denaturing polyacrylamide gel at 200–300 V for 2–3 h.

2 Stain the gel in EtBr for 10 min and view on a UV transilluminator.

3 Excise a gel slice encompassing the size of RNA (3'-adaptor + small RNA + 5'-adaptor) and crush into pieces. Elute RNA from the gel pieces by immersion in 0.3 M NaCl at 4°C overnight.

4 Precipitate RNA from the aqueous phase in 1/10th Vol. NaOAc, 2.5 Vol. ethanol and glycogen at –20°C for more than 1 hr.

5 Centrifuge to pellet the RNA and wash in 80% ethanol; after 5 min air dry under a laminar hood; resuspend the pellet in 20 μl DEPC-treated H_2O. Consider this the stock, and use 8 μl for the next step.

3.9. Reverse Transcription to Generate cDNA and PCR for Further Amplification

A 10-μl reaction containing 8 μl adaptor-ligated small RNA, 1 μl reverse primer with the sequence complementary to a 3' adaptor and 1 μl (10 mM) dNTP mix is incubated at 65°C for 10 min and quick chilled on ice. cDNA synthesis is carried out in a 20 μl reaction containing 4 μl 5X RT buffer, 2 μl 0.1 M DTT, 1 μl Superscript II, 1 μl RNaseOUT, including the above primer-annealed RNA mixture. Mix well by pipetting up and down and apply a short spin. Incubate the reactions at 42°C for 1 h. Stop the reaction by incubating at 70°C for 15 min, then snap cool on ice; after a brief spin, store at –20°C until use (consider this a second stock and use 4 μl for the next step).

PCR reaction is carried out in a 50-μl reaction containing 10 μl 5X buffer, 4 μl RT reaction mixture, 2 μl dNTP mix (2 mM), 1 μl primer1 (10 μM), 1 μl primer2 (10 μM), 31.5 μl ddH_2O and 0.5 μl Phusion DNA polymerase. Here we recommend the use of Phusion DNA polymerase because of its high-fidelity proofreading property. PCR is performed at 98°C for 1 min, then 15 cycles of 98°C for 10 s, 60°C for 10 s and 72°C for 30 s, with a final extension at 72°C for 5 min.

3.10. PCR Product Purification

1. Run the entire sample on a 12% native polyacrylamide gel (5 ml 10X TBE, 30 ml 40% acrylamide-bisacrylamide solution, 63.3 ml H_2O, 700 μl 10% APS and 37 μl TEMED) and in parallel run a 10-bp ladder at 200–300 V for about 2–3 h.
2. Stain the gel in EtBr for 5–10 min and visualize under a UV transilluminator.
3. Cut out a gel slice corresponding to the length of small RNA plus 5' and 3' adaptors.

4. To elute DNA, immerse gel pieces in 0.3 M NaCl and incubate at 4°C overnight.
5. To precipitate DNA, add 1/10th Vol. NaOAc, 2.5 Vol. ethanol and glycogen and leave at –20°C for more than 1 h. Centrifuge to pellet the DNA, and after an 80% ethanol wash, dry the pellet and resuspend in 20 μl dH_2O.

3.11. Quality Check

The quality of the library needs to be checked by conventional cloning and sequencing prior to high-throughput sequencing.

Phusion is a high-fidelity proofreading enzyme and yields blunt-ended PCR products. Taq DNA polymerase treatment is necessary to add A-residue at the 3' end of the PCR products to facilitate cloning into a TOPO-TA vector, which has T-residue overhangs.

1. Add "A" to the 3' end of the PCR product: 2 μl gel-eluted DNA, 1 μl 10X buffer, 1 μl dATP mix (2 mM), 0.5 μl Taq DNA polymerase; incubate the reaction at 72°C for 20 min.
2. Clone PCR fragments into TOPO-TA vector (PCR2.1 Vector, Invitrogen): 4.5 μl fresh Taq-treated products, 1 μl salt solution, and 0.5 μl TOPO vector to a total Vol. of 6 μl; incubate the reaction at RT for 10 min. Transform the ligation product into Top10 competent cells. Spread 10–50 μl of each transformation mixture onto LB plates containing 50 μg/ml kanamycin and X-gal. Incubate overnight at 37°C. The next day, pick up the white colonies and test for presence of inserts by colony-PCR and size determination by agarose gel electrophoresis. A blue colony is used for negative control.
3. Colony PCR: reaction is carried out by adding 2.0 μl 10X PCR buffer, 0.5 μl dNTP mix, 1.0 μl M13F primer, 1.0 μl M13R primer, 0.5 μl Taq polymerase, 15.0 μl H_2O. Choose individual colonies and put onto a reference plate, then into individual tubes containing PCR mix. Cycles are (1) 94°C for 2 min, (2) 94°C for 30 s, (3) 50°C for 30 s, (4) 72°C for 30 s; repeat (2) to (4) for 25 cycles; (5) 72°C for 5 min and (6). After PCR is finished, use 8 μl of the reaction to check on a 1.5% agarose gel.
4. Sequencing: the PCR products longer than those from empty vector (blue colony) are subjected to conventional sequencing with \50 clones per library. A BLAST search can identify the sequences of small RNAs. If the small RNAs identified have the correct length of 20–30 nt and no self-ligation clones, the DNA product can be directly used for high-throughput parallel sequencing.

3.12. 5'-P-dependent Cloning: 5'-adaptor ligation

This method ligates the 5' adaptor directly onto gel-fractionated small RNAs. A 30-μl reaction containing 20 μl purified small RNA, 3 μl 10X buffer, 2 μl T4 RNA ligase (5 U/μl), 2 μl 5' adaptor (25 μM), 2 μl H_2O and 1 μl RNaseOUT is mixed gently by tapping. After a brief spin, incubate the tube at RT for 6 h. Run the reaction onto a denaturing polyacrylamide gel; excise a gel piece with the correct size of RNA product, elute and dissolve in DEPC-treated 20 μl water as described above. The product is used directly for 3' adaptor ligation.

3.13. 3' Adaptor Ligation

3' Adaptor ligation of small RNA is carried out in a 30 μl reaction containing 20 μl purified 5' adaptor-ligated small RNA, 3 μl 10X buffer, 2 μl T4 RNA ligase (5 U/μl), 2 μl 3' adaptor (25 μM), 2 μl H_2O, and 1 μl RNaseOUT, then mixed gently by tapping; after a brief spin, incubate the tube at RT for 6 h. The 3' adaptor should also have the inverted nucleotide at the end to avoid self-ligation. The subsequent steps are the same as for the 5' P-independent method. The difference between these two methods is dephosphorylation prior to 3' adaptor ligation and phosphorylation prior to 5' adaptor ligation in the 5' P-independent method (**Fig. 11.1**).

3.14. High-Throughput Parallel Sequencing for Small-RNA Identification

The constructed small RNA libraries can be subjected to high-throughput sequencing to obtain highly enriched sequence information. Instead of cloning the small RNA libraries into plasmid vectors for conventional sequencing, high-throughput parallel sequencing approaches directly amplify and sequence small RNA libraries on microbeads and yield hundreds of thousands of sequence reads simultaneously. High-throughput sequencing techniques, including MPSS *(21)* (www.solexa.com) and 454 sequencing (454 Life Sciences, www.454.com) *(22)*, have been demonstrated to be robust approaches for small RNA discovery *(11, 13–18, 23)*. However, MPSS yields only 17 nt of sequence and requires further analysis to gain the sequence information for the full-length small RNAs; 454 sequencing involves generation and detection of pyrophosphate signals and photons, and can generate more than 100-nt reads. It can sequence 25 million bases with 99% accuracy in a 1-h run *(22)* and is ideal for small-RNA sequencing. To replace MPSS, Solexa, Inc. (Illumina) recently developed a four-color DNA sequencing-by-synthesis (SBS) approach, which can generate more than 30 million 30- to 35-nt reads per run with high accuracy and is applicable for small-RNA discovery. Applied Biosystems also developed a ligation-based SOLiD high-throughput sequencing system, which can generate more than 1 gigabase data of 25- to 35-nt reads. The comparison among these three technologies is summarized in **Table 11.1.** These high-throughput sequencing technologies not only provide rich sequence information but also can be used for global expression analysis of protein coding genes or small RNAs. The small-RNA libraries constructed as described above can be sequenced using

Table 11.1
1High-throughput sequencing approaches

	454	SBS	SOLiD
Reaction	Emulsified PCR in a water-in-oil mixture	Solid-phase bridge amplification	Emulsion PCR ligation based sequencing
Signal	Pyrophosphate and photons	Fluorescent signal	Fluorescent signal
Reaction base	Micro-beads within a picotiter plate	Solid surface of a flow cell	Micro-beads
Read Length	About 100–300 nt	About 30–35 nt	About 25–35 nt
Data size	> 0.5 Gigabase/run	>1 Gigabase/run	>1 Gigabase/run
Websites	www.454.com	www.illumina.com	www.appliedbiosystems.com

either of these technologies to generate rich resources for small RNA identification and global expression analysis.

4. Notes

1. All required materials should be autoclaved prior to use.
2. All autoclaved materials, baked glassware, and solutions should be handled with gloved hands to avoid RNase contamination.
3. Tissue must be frozen immediately in liquid N (delayed freezing might cause RNA degradation and loss of yield).
4. Tissue must be ground to a fine powder to avoid reduced RNA yield.
5. Trizol can be added first to the fine powder, then the frozen tissue can be thawed in the presence of Trizol to avoid degradation, and chloroform added and mixed well; then, contents can be transferred to a polyprophane tube (phenol resistant), because the next step requires high-speed centrifugation.
6. The Trizol extraction step requires centrifugation at room temperature.
7. Dissolve the RNA pellet in 50% formamide for gel analysis or in DEPC-treated water for enzymatic reactions.
8. If the pellet is difficult to dissolve, try heating at 65°C for 5 min.
9. In general, add enough Trizol/chloroform to avoid increased protein contamination in the extraction. Sometimes, RNA

pellet derived from crop plant tissues appears brown or even black after precipitation, this is because some secondary metabolites or polysaccharides accumulated in the sample; this could be avoided by adding more Trizol/chloroform. In pure and high quality RNA preparation, the pellet appears white.

Acknowledgments

Research in Jin Laboratory is supported by NSF Career Award MCB-0642843 and a University of California Discovery Grant Bio06-10566 and AES-CE Research Allocation Award PPA-7517H.

References

1. Hamilton, A.J. and Baulcombe, D.C. (1999) A species of small antisense RNA in posttranscriptional gene silencing in plants. *Science*, **286**, 950–952.
2. Lee, R.C., Feinbaum, R.L. and Ambrose, V. (1993) The C.elegans heterochronic gene lin-4 encodes small RNAs with antisense complementarity to lin-4. *Cell*, **75**, 843–854.
3. Valencia-Sanchez, M.A., Liu, J., Hannon, G.J. and Parker, R. (2006) Control of translation and mRNA degradation by miRNAs and siRNAs. *Genes Dev*, **20**, 515–524.
4. Carrington, J.C. and Ambros, V. (2003) Role of microRNAs in plant and animal development. *Science*, **301**, 336–338.
5. Baulcombe, D. (2004) RNA silencing in plants. *Nature*, **431**, 356–363.
6. Bartel, D.P. (2004) MicroRNAs: genomics, biogenesis, mechanism, and function. *Cell*, **116**, 281–297.
7. Reinhart, B.J., Weinstein, E.G., Rhoades, M.W., Bartel, B. and Bartel, D.P. (2002) MicroRNAs in plants. *Genes Dev*, **16**, 1616–1626.
8. Allen, E., Xie, Z., Gustafson, A.M. and Carrington, J.C. (2005) microRNA-directed phasing during trans-acting siRNA biogenesis in plants. *Cell*, **121**, 207–221.
9. Katiyar-Agarwal, S., Morgan, R., Dahlbeck, D., Borsani, O., Villegas, A.J., Zhu, J.K., Staskawicz, B.J. and Jin, H. (2006) A pathogen-inducible endogenous siRNA in plant immunity. *Proc Natl Acad Sci USA*, **103**, 18002–18007.
10. Borsani, O., Zhu, J., Verslues, P.E., Sunkar, R. and Zhu, J.K. (2005) Endogenous siRNAs derived from a pair of natural cis-antisense transcripts regulate salt tolerance in Arabidopsis. *Cell*, **123**, 1279–1231.
11. Henderson, I.R., Zhang, X., Lu, C., Johnson, L., Meyers, B.C., Green, P.J. and Jacobsen, S.E. (2006) Dissecting Arabidopsis thaliana DICER function in small RNA processing, gene silencing and DNA methylation patterning. *Nat Genet*, **38**, 721–725.
12. Lu, C., Tej, S.S., Luo, S., Haudenschild, C.D., Mayers, B.C. and Green, P.J., (2005) Elucidation of the small RNA component of the transcriptome. Science, **309**, 1567–1579.
13. Qi, Y., He, X., Wang, X.J., Kohany, O., Jurka, J. and Hannon, G.J. (2006) Distinct catalytic and non-catalytic roles of ARGONAUTE4 in RNA-directed DNA methylation. *Nature*, **443**, 1008–1012.
14. Zhang, X., Henderson, I.R., Lu, C., Green, P.J. and Jacobsen, S.E. (2007) Role of RNA polymerase IV in plant small RNA metabolism. *Proc Natl Acad Sci USA*, **104**, 4536–4541.
15. Howell, M.D., Fahlgren, N., Chapman, E.J., Cumbie, J.S., Sullivan, C.M., Givan, S.A., Kasschau, K.D. and Carrington, J.C. (2007) Genome-wide analysis of the RNA-DEPENDENT RNA POLYMERASE6/DICER-LIKE4 pathway in Arabidopsis reveals dependency on miRNA- and tasiRNA-directed targeting. *Plant Cell*, **19**, 926–942.

16. Fahlgren, N., Howell, M.D., Kasschau, K.D., Chapman, E.J., Sullivan, C.M., Cumbie, J.S., Givan, S.A., Law, T.F., Grant, S.R., Dangl, J.L. et al. (2007) High-throughput sequencing of Arabidopsis microRNAs: evidence for frequent birth and death of MIRNA genes. *PLoS ONE*, **2,** e219.
17. Kasschau, K.D., Fahlgren, N., Chapman, E.J., Sullivan, C.M., Cumbie, J.S., Givan, S.A. and Carrington, J.C. (2007) Genome-wide profiling and analysis of Arabidopsis siRNAs. *PLoS Biol*, **5(3)**, e57.
18. Rajagopalan, R., Vaucheret, H., Trejo, J. and Bartel, D.P. (2006) A diverse and evolutionarily fluid set of microRNAs in Arabidopsis thaliana. *Genes Dev*, **20**, 3407–3425.
19. Pak, J. and Fire, A. (2007) Distinct populations of primary and secondary effectors during RNAi in C.elegans. *Science*, **315,** 241–244.
20. Pfeffer, S., Sewer, A., Lagos-Quintana, M., Sheridan, R., Sander, C., Grasser, F.A., van-Dyk, L.F., Ho, C.K., Shuman, S., Chien, M. et al. (2005) Identification of microRNAs of the herpesvirus family. *Nat Methods*, **4,** 269–276.
21. Brenner, S., Johnson, M., Bridgham, J., Golda, G., Lloyd, D.H., Johnson, D., Luo, S., McCurdy, S., Foy, M., Ewan, M. et al. (2000) Gene expression analysis by massively parallel signature sequencing (MPSS) on microbead arrays. *Nat Biotechnology*, **6,** 630–634.
22. Margulies, M., Egholm, M., Altman, W.E., Attiya, S., Bader, J.S., Bemben, L.A., Berka, J., Braverman, M.S., Chen, Y.J., Chen, Z. et al. (2005) Genome sequencing in microfabricated high-density picolitre reactors. *Nature*, **437,** 376–380.
23. Lu, S.W., Sun, Y.-H., Shi, R., Clark, C., Li, L. and Chiang, V.L. (2005) Novel and mechanical stress–responsive microRNAs in Populus trichocarpa that are absent from Arabidopsis. *The Plant Cell*, **17,** 2186–2203.

Chapter 12

The Use of Chemical Genomics to Investigate Pathways Intersecting Auxin-Dependent Responses and Endomembrane Trafficking in *Arabidopsis Thaliana*

Lorena Norambuena, Glenn R. Hicks and Natasha V. Raikhel

Abstract

Plant endomembrane system is essential for viability and necessary for proper development and signal transduction signal processes. Links between the endomembrane system and auxin signaling have been reported by classical genetics screens. However, the relationship between these processes is not well understood. Chemical genomics is a powerful approach to dissect various processes overcoming lethality and redundancy issues. This approach uses small molecules to modify or disrupt the function of specific proteins and biological processes. We present a screen in *Arabidopsis thaliana* to identify compound affecting auxin-dependent responses and components of the endomembrane system. A gravitropic-response based screen is performed in Arabidopsis seedlings. The identified gravitropic effectors are tested in terms of auxin responsiveness and their effects on endomembrane compartments. These bioactive compounds will be valuable tools for dissecting endomembrane trafficking and auxin signaling processes.

Key words: Plant, endomembrane system, chemical genomics, gravitropism, auxin signaling.

1. Introduction

The plant endomembrane system is critical for biosynthetic and endocytic trafficking. Recently, links between the endomembrane system, plant signaling and development have emerged. These links include abscisic acid and auxin signaling, tropic responses, and pathogen defense *(1, 2)*. Classical mutant screens have shown that the endomembrane system is intimately involved in auxin-dependent responses such as the gravitropic response. For instance, several mutants in genes that encode SNARE (*S*oluble *N*-ethylmaleimide-sensitive *a*daptor protein *re*ceptor) proteins

Sean Cutler, Dario Bonetta (eds.), *Plant Hormones: Methods and Protocols, Second Edition, vol. 495*

DOI 10.1007/978-1-59745-477-3_12 Springerprotocols.com

are agravitropic and have defective vacuoles *(3, 4)*. Furthermore, polar auxin transport is dependent upon components of the endomembrane system. Auxin transporters such as the PIN proteins display dynamic cycling between plasma membrane and endosomal compartments *(5–7)*. Recently, it has been shown that PIN proteins are internalized by a clathrin dependent mechanism *(8)*.

Chemical genomics (i.e., genomics-scale chemical genetics) is the use of small molecules to modify or disrupt the function of specific proteins and modify biological processes *(9–11)*. At a practical level, chemical genomics entails screening diverse chemical libraries for effectors of a process of interest. This approach overcomes the limitations of mutant lethality because chemicals can be applied in a spatial or temporal manner, are tunable, and their effects are frequently reversible. In forward chemical genomic screens, large numbers of compounds are tested for their ability to alter a specific phenotype. Therefore it is crucial to establish a straightforward and robust phenotypic screen that will permit the testing of thousands of chemicals in a high-throughput manner. As a template for laboratories interested in adopting chemical genomics, we present an approach to identify bioactive chemicals that affect both auxin-dependent responses and components of the endomembrane system. A screening based upon the gravitropic response in *Arabidopsis thaliana* is performed to identify chemical effectors. Then, their impact on the morphology and targeting to endomembrane compartments is tested using GFP fusion markers.

2. Materials

2.1. Plant Growth

1. Seeds are sterilized and stratified in darkness for 48 h at 4°C prior to plating. They are germinated and grown in an incubator at 22°C under standard conditions of humidity and photoperiod appropriate for growing Arabidopsis.
2. The culture media for making plates is 0.5 × Murashige and Skoog (MS) media (PlantMedia, Dublin, OH) pH 5.6 containing 2% sucrose and 0.3% GELRITE (RPI, Illinois, IL).

2.2. Chemical Treatments

The chemical library used in this screen is the DIVERSet library (ChemBridge, San Diego, CA). This library contains 10,000 small organic molecules in a 96-well format plate. In order to prepare master plates with stock solutions, DMSO (Fischer Scientific) is used as solvent.

1. Dissolve 0.1 mg of each compound in 20 μl of 100% DMSO for master plates and store at –80°C.

2. Prepare working solution plates by diluting each master plate five times with water in polypropylene 96 well plates (Corning, # 3355). The final concentration of the compound solutions will range between 2–4 mM (depending upon the mass of the compounds) in 20% DMSO.
3. Re-array the working solution library to a 24-well format by using a Bio-Tek (Winooski, VT) Precision 2000 liquid-handling robot or similar fluid-handling robot. Although more tedious, this can also be done using a hand-held multi-channel pipetter. The working solution plates are stored at –80°C.
4. For the primary screen:
   - 4.1. Add 10 μl of each chemical to each well of a 24-well plate (Corning, #3526).
   - 4.2. To each well, add 390 μl of MS agar (at 50°C) and mix it by shaking the plate gently. Allow the agar to solidify for 30–40 min. The final concentration of compound is 25 μg/ml in 0.5% DMSO. The molarities depend upon compound mass and can be calculated for individual chemicals.
5. For the secondary screen: Prepare in 24-well plates an array of MS media containing different concentrations (0, 0.25, 0.5, 1, 5, and 10 μg/ml) of each compound identified from the primary screen. For a screen such as this, based upon alterations in gravitropic responses, typically about 1–5% of the compounds will appear positive in the primary screen.

2.3. Characterization of the Gravitropic Effectors

2.3.1. Quantification of Gravitropic Response

1. Prepare on 4-well plate (Nalge Nunc, Model 176597) an array of MS media containing different concentrations (0, 0.25, 0.5, 1, 5 and 10 μg/ml) of each bioactive compound. The recommended final volume for each well is 4.5–6.0 ml. These plates will be used to quantify gravitropic responses.

2.3.2. Auxin Responsiveness

1. Sow seed of an Arabidopsis line expressing the auxin-responsive reporter gene DR5::GUS *(12)* on MS media (without chemicals) and incubate under normal conditions. Incubate the plates vertically to allow seedlings grow straight across the surface of the media.
2. Prepare 10 ml of MS media agar containing 10 μM indole-3 acetic acid (IAA) or other auxins such as NAA or 2,4D in a 15 ml tube (Corning, 430790). Store it in darkness. This will be used for auxin transport assay.

3. GUS staining solution: 1 mg/ml X-Gluc (5-bromo-4-chloro-3-indolyl β-D-glucuronide cyclohexamine salt); 100 mM $NaHPO_4$ Buffer (pH 7.2); 10 mM EDTA (pH 8); 0.1% Triton X-100; 2 mM Potassium Ferrocyanide; 2 mM Potassium Ferricyanide. It is convenient to first dissolve the X-Gluc in dimethylformamide (DMF; Fischer Scientific) at 100 μg/ml. GUS staining solution can be made and stored at –20°C for at least one year.

2.4. Effect of Gravitropic Effectors on Endomembrane System Compartments

Prepare on twenty four-well plate, an array of MS media containing at least three different concentrations (1, 5 and 10 μg/ml) of each bioactive compound. These plates will be used to analyze the effect of the bioactive compounds on compartments of the endomembrane system.

3. Methods

The methods described below outline (i) a screen of a small synthetic organic chemical library based upon the gravitropic response of Arabidopsis seedlings (ii) the characterization of the effectors in terms of auxin responses, and (iii) the effects of the chemicals on morphology and targeting to the endomembrane system compartments.

3.1. Screening a Chemical Library for Effectors of Gravitropic Response

A primary screen is described using a 10,000 compound library to identify chemicals that affect the gravitropic response of Arabidopsis seedlings. In order to confirm the results from primary screening, the effect of the chemicals is re-tested at different concentrations in a second screen.

3.1.1. Standard Gravitropism Assay

1. *Arabidopsis thaliana* (ecotype Columbia- 0) is used for the gravitropic screen. Place 5–8 stratified Arabidopsis seeds towards lower half of the well of 24-well plate (when placed vertically) containing media supplemented with chemicals.
2. Incubate the plates in light for 4–14 h to induce germination at 22°C.
3. Incubate plates vertically in darkness at 22°C for 48 h.
4. Turn the plates 90° and incubate them for 24–72 h.
5. Image the plates by placing them on a flat-bed scanner (Model 2450 scanner; Epson, Long Beach, CA) to score and record the response of the seedlings.

3.1.2. Gravitropic Screen

1. Primary screen: 10,000 compounds are tested at 25 μg/ml in 0.5% DMSO on 24-well plate format (2.2 primary screen) (**Fig. 12.1A**). A chemical is considered as a putative positive when the majority of seedling stems in a well fails to bends or display an enhanced gravitropic response.

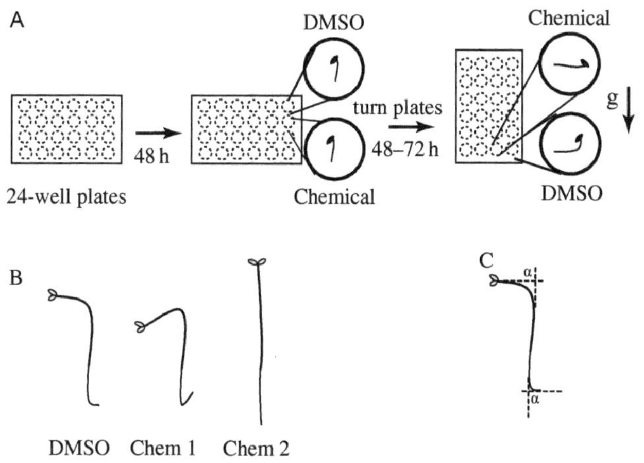

Fig. 12.1. (**A**) Screen for chemicals that affect gravitropism. The chemical library was screened in a 24-well format, and seedlings were scored for gravitropic response after reorientation. Chemicals dissolved in 20% DMSO were added to wells. Control wells contained an equivalent concentration of the solvent. The gravity vector (g) is indicated by an arrow (on the *right*). Reproduced from reference 12 with permission of PNAS, Copyright 2005 National Academy of Sciences, USA. (**B**) Representation of expected chemical screening results. (**C**) Root and hypocotyl curvature angles (α).

2. Secondary screen: The gravitropism assay is performed under the same conditions as the primary screen, but at concentrations of 0, 0.25, 5 and 10 μg/ml in a 24-well plate format (2.2 secondary screen). A chemical is considered as a confirmed positive if (1) the inhibition or the enhancement of the gravitropic response is present in the re-test and (2) the effect of the chemical appears to be dose-dependent.

3.2. Characterizing the Gravitropic Response Effectors

3.2.1. Quantification of Gravitropic Response

1. Sow 25–30 seeds in a well of a 4-well plate containing different concentrations of each confirmed positive. The seeds are placed in the middle of the well allowing the analysis of both root and hypocotyl responses. The chemical concentrations are 0, 0.25, 0.5, 1, 5 and 10 μg/ml (**Section 2.3.1**).
2. Perform the gravitropic assay as described in **Section 3.1.1** (*Standard gravitropism assay*).
3. Scan the plates. Figure 1B shows examples of the phenotypes that may be observed including both inhibition and enhancement of the gravitropic response.
4. Measure the curvature angles of roots and hypocotyls from the vertical using the angle tool in SCION IMAGE (available free on-line from Scion, Frederick, MD) or similar image analysis software. The angles to be measured are indicated in **Fig. 12.1C**.

3.2.2. Auxin Responsiveness

The transgenic Arabidopsis line expressing β-glucuronidase (GUS) protein under the control of the auxin-inducible promoter DR5 (DR5::GUS) is used to examine the effect of the bioactive chemicals on auxin signaling *(12)*. The ability to induce the DR5 promoter is analyzed to test whether the chemical is acting as an auxin. In addition, the effects of compounds on auxin transport and signaling are tested.

3.2.2.1. DR5 Promoter Activity Induction

1. Transfer four to six DR5::GUS seedlings (7-days old) to media containing 0, 1, 5 and 10 μg/ml of compound.
2. Incubate seedlings in the presence of compound for 5 h in an incubator at 22°C.
3. β-glucuronidase (GUS) activity is determined as described below (**Section 3.2.2.4**).

3.2.2.2. Auxin Transport Assay

1. Transfer four to six DR5::GUS seedlings (7-days old) to media containing a compound of interest (0, 1, 5 and 10 μg/ml).
2. Plates are incubated for 14 h in an incubator at 22°C.
3. Place a 1 mm block of MS media agar containing 10 μM indole-3 acetic acid (IAA) at the tip of the root. Use a glass Pasteur pipette to remove a core of IAA/MS agar from the 15 ml tube of **Section 2.3.2**, pull it out with a syringe or bulb.
4. Incubate the plates for additional 5 h in the incubator at 22°C.
5. Measure GUS activity.

3.2.2.3. Induction of DR5 Promoter by Auxin

1. Transfer four or six 7 day-old DR5::GUS seedlings to media containing the compound of interest at 0, 1, 5 and 10 μg/ml.
2. Incubate the plates for 14 h into the incubator at 22°C.
3. Transfer the seedlings to media containing 0 or 1 μM IAA and 0, 1, 5 or 10 μg/ml of compound.
4. Incubate the plates for 5 more hours in the incubator at 22°C.
5. Measure GUS activity.

3.2.2.4. GUS Activity Staining

1. Harvest the tissue directly into the GUS staining solution.
2. Place the plate in a vacuum desiccator and draw a vacuum for 15 min. Release the vacuum slowly to avoid cell rupture.
3. Incubate the staining reaction in darkness at 37°C for 4–8 h. Wrap the plate with plastic and foil paper.
4. Remove the solution and replace with 70% ethanol (EtOH, Fisher Scientific).
5. Change the EtOH solution as much as necessary to remove the chlorophyll from cotyledons.
6. Analyze and image the stained seedlings using bright-field compound microscopy.

3.3. Effect of the Gravitropic Effectors on Endomembrane System Compartments

The effect of the bioactive compounds on morphology and targeting to compartments of the endomembrane system can be monitored using fluorescent marker proteins. **Table 12.1** shows a list of either GFP or YFP fusion markers for different compartments and **Fig. 12.2** shows representative phenotypes observed using these markers with hit compounds.

1. Sow seeds from transgenic reporter lines of interest on MS agar containing different concentrations of compound (0, 1, 5 and 10 μg/ml; **Section 2.4**).

Table 12.1
Fluorescent fusion markers of endomembrane compartments and cytoskeleton

Fluorescent marker	Subcellular localization	Reference
GFP:KDEL	Endoplasmic reticulum	*(14)*
GFP-NAG1	Golgi apparatus	*(15)*
MAN-GFP	Golgi apparatus	*(16)*
TLG2a-GFP	Trans Golgi network	*(17)*
VHA-a1-GFP	Trans Golgi network	*(18)*
ARA6-GFP	Endosomes	*(15)*
ARA7-GFP	Endosomes	*(17)*
RFP-ARA7	Endosomes	*(19)*
SNX1-GFP	Endosomes	*(17)*
BOR1-GFP	Endosomes	*(19)*
GFP-δ-TIP	Tonoplast	*(20)*
VHAa2-GFP	Tonoplast	*(18)*
VHAa3-GFP	Tonoplast	*(18)*
Ale-GFP	Lytic vacuole	*(21)*
GFP-CHI	Protein storage vacuole	*(21)*
PIP2A-YFP	Plasma membrane	*(22)*
PHOT1-GFP	Plasma membrane	*(23)*
GFP-MAP4	Microtubule cytoskeleton	*(24)*
ABD2-GFP	Actin cytoskeleton	*(25)*

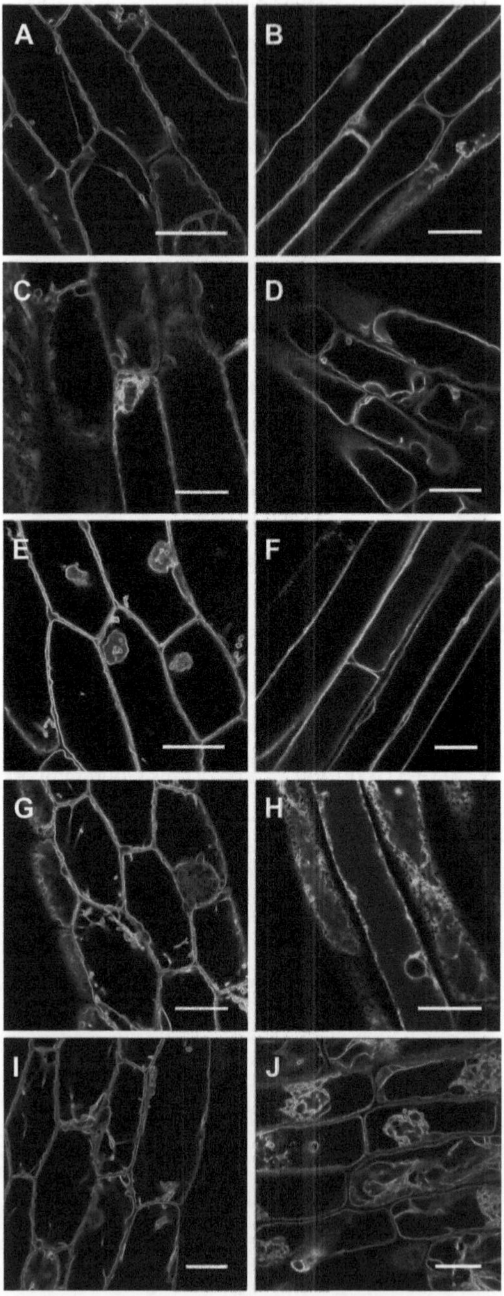

Fig. 12.2. Four compounds affect vacuole morphology or targeting of the reporter molecule to the tonoplast in GFP:δ-TIP seedlings. Hypocotyls (**A, C, E, G,** and **I**) and roots (**B, D, F, H,** and **J**) were examined by laser-scanning confocal microscopy after germination and growth on the chemicals for 7 days in the light. Chemicals: (**C** and **D**) 5403629 at 6.6 μM, (**E** and **F**) 5271050 at 3.7 μM, (**G** and **H**) 5850247 at 35 μM, and (**I** and **J**) 6220480 at 30 μM final concentration. Chemical 5403629 was inhibitory to growth at the concentration tested; so seedlings were grown on medium without the chemical for 4 days and then treated with the compound for 48 h. The arrows in **C** and **D** indicate vesiculations; **E**, aggregates; **H**, endoplasmic reticulum patterning; and **J**, aggregates. (**A** and **B**) Images of a control seedling grown in the presence of a concentration of DMSO equivalent to that of treated seedlings are shown for comparison (*bars*, 20 μm). Reproduced from reference *(12)* with permission of PNAS, Copyright 2005 National Academy of Sciences, USA.

2. Incubate the plates horizontally for 7 days at 22°C (normal conditions).
3. Pull out the seedling carefully and mount it on a slide (Gold Seal, Portsmouth, NH, US. #3011) with water and cover it with a coverglass (VWR International, #48393172.).
4. Analyze and image the different tissues (cotyledons, hypocotyl and root) by confocal microscopy (Leica TCS SP2/UV, Wetlzer, Germany) noting the morphology of the compartments in the presence and absence of compound.

4. Notes

1. The screen and all experiments have to include control plates without chemicals containing 0.5% DMSO.
2. To avoid bacteria or fungal contamination, the gravitropic screen may be done with MS media without sucrose.
3. Some chemicals may affect Arabidopsis growth. To assess the affects on hypocotyl and root growth, lengths can be measured with the SCION IMAGE multi-line tool and expressed as a percentage of control seedling growth. A multi-line tool facilitates the approximate measurement of lengths even in the presence of curvature.
4. The quantification of gravitropic response must be done at least in duplicate. The data should be graphed (degree of curvature vs chemical concentration) as the average of 30–50 seedlings total per data point with standard error.
5. For chemicals that inhibit the gravitropic response, the concentrations that result in half-maximal inhibition (IC_{50}s) of the curvature response, can be derived from the inhibition curves.
6. The effects of chemicals may be tested in terms of inducibility and reversibility.
    6.1. For inducibility: seedlings are germinated in the absence of chemical and then transferred to a plate with chemical to test for an altered phenotype due to a bioactive chemical.
    6.2. For reversibility: seedlings are germinated in the presence of the chemical. Then they are transferred to a plate without chemical to test for recovery of wild type phenotype indicating that the chemical effect is reversible.
7. The inhibitor 2,3,5-triiodobenzoic acid (TIBA) can be included as a positive control of inhibition in the auxin transport assay.
8. For confocal microscopic imaging at a high numerical aperture (1.2–1.3), a water immersion objective (63x) is employed. A 488-nm or a 514-nm laser line from an argon ion laser is used to excite GFP and YFP, respectively.

9. A chemical genomic screen has a hit rate approximately 0.01–0.1% (confirmed after secondary screen). A screening using the Diverset library had a 0.05% hit rate of chemicals that affect both gravitropic and the endomembrane system *(13)*.

Acknowledgments

This work was supported by a grant from the National Science Foundation (MCB-0515963).

References

1. Surpin, M., and Raikhel, N. (2004) Traffic jams affect plant development and signal transduction. *Nat Rev Mol Cell Biol* **5**, 100–109
2. Carter, C. J., Bednarek, S. Y., and Raikhel, N. V. (2004) Membrane trafficking in plants: new discoveries and approaches. *Curr Opin Plant Biol* 7, 701–707
3. Morita, M. T., Kato, T., Nagafusa, K., Saito, C., Ueda, T., Nakano, A., and Tasaka, M. (2002) Involvement of the Vacuoles of the Endodermis in the Early Process of Shoot Gravitropism in Arabidopsis. *Plant Cell* **14**, 47–56
4. Yano, D., Sato, M., Saito, C., Sato, M. H., Morita, M. T., and Tasaka, M. (2003) A SNARE complex containing SGR3/AtVAM3 and ZIG/VTI11 in gravity-sensing cells is important for Arabidopsis shoot gravitropism. *Proc Natl Acad Sci USA* **100**, 8589–8594
5. Geldner, N., Friml, J., Stierhof, Y. D., Jurgens, G., and Palme, K. (2001) Auxin transport inhibitors block PIN1 cycling and vesicle trafficking. *Nature* **413**, 425–428
6. Paciorek, T., Zazimalova, E., Ruthardt, N., Petrasek, J., Stierhof, Y. D., Kleine-Vehn, J., Morris, D. A., Emans, N., Jurgens, G., Geldner, N., and Friml, J. (2005) Auxin inhibits endocytosis and promotes its own efflux from cells. *Nature* **435**, 1251–1256
7. Boutte, Y., Crosnier, M. T., Carraro, N., Traas, J., and Satiat-Jeunemaitre, B. (2006) The plasma membrane recycling pathway and cell polarity in plants: studies on PIN proteins. *J Cell Sci* **119**, 1255–1265
8. Dhonukshe, P., Aniento, F., Hwang, I., Robinson, D. G., Mravec, J., Stierhof, Y. D., and Friml, J. (2007) Clathrin-mediated constitutive endocytosis of PIN auxin efflux carriers in Arabidopsis. *Curr Biol* **17**, 520–527
9. Stockwell, B. R. (2000) Chemical genetics: ligand-based discovery of gene function. *Nat Rev Genet* **1**, 116–125
10. Yeh, J. R., and Crews, C. M. (2003) Chemical genetics: adding to the developmental biology toolbox. *Dev Cell* **5**, 11–19
11. Blackwell, H. E., and Zhao, Y. (2003) Chemical Genetic Approaches to Plant Biology. *Plant Physiol* **133**, 448–455
12. Ulmasov, T., Murfett, J., Hagen, G., and Guilfoyle, T. J. (1997) Aux/IAA Proteins Repress Expression of Reporter Genes Containing Natural and Highly Active Synthetic Auxin Response Elements. *Plant Cell* **9**, 1963–1971
13. Surpin, M., Rojas-Pierce, M., Carter, C., Hicks, G. R., Vasquez, J., and Raikhel, N. V. (2005) The power of chemical genomics to study the link between endomembrane system components and the gravitropic response. *Proc Natl Acad Sci USA* **102**, 4902–4907
14. Boevink, P., Oparka, K., Santa Cruz, S., Martin, B., Betteridge, A., and Hawes, C. (1998) Stacks on tracks: the plant Golgi apparatus traffics on an actin/ER network. *Plant J* **15**, 441–447
15. Grebe, M., Xu, J., Mobius, W., Ueda, T., Nakano, A., Geuze, H. J., Rook, M. B., and Scheres, B. (2003) Arabidopsis sterol endocytosis involves actin-mediated trafficking via ARA6-positive early endosomes. *Curr Biol* **13**, 1378–1387
16. Nebenfuhr, A., Gallagher, L. A., Dunahay, T. G., Frohlick, J. A., Mazurkiewicz, A. M., Meehl, J. B., and Staehelin, L. A. (1999)

Stop-and-Go Movements of Plant Golgi Stacks Are Mediated by the Acto-Myosin System. *Plant Physiol* **121**, 1127–1141
17. Jaillais, Y., Fobis-Loisy, I., Miege, C., Rollin, C., and Gaude, T. (2006) AtSNX1 defines an endosome for auxin-carrier trafficking in Arabidopsis. *Nature* **443**, 106–109
18. Dettmer, J., Hong-Hermesdorf, A., Stierhof, Y.-D., and Schumacher, K. (2006) Vacuolar H+-ATPase Activity Is Required for Endocytic and Secretory Trafficking in Arabidopsis. *Plant Cell* **18**, 715–730
19. Takano, J., Miwa, K., Yuan, L., von Wiren, N., and Fujiwara, T. (2005) Endocytosis and degradation of BOR1, a boron transporter of Arabidopsis thaliana, regulated by boron availability. *Proc Natl Acad Sci USA* **102**, 12276–12281
20. Cutler, S. R., Ehrhardt, D. W., Griffitts, J. S., and Somerville, C. R. (2000) Random GFP::cDNA fusions enable visualization of subcellular structures in cells of Arabidopsis at a high frequency. *Proc Natl Acad Sci USA* **97**, 3718–3723
21. Fluckiger, R., De Caroli, M., Piro, G., Dalessandro, G., Neuhaus, J.-M., and Di Sansebastiano, G.-P. (2003) Vacuolar system distribution in Arabidopsis tissues, visualized using GFP fusion proteins. *J Exp Bot* **54**, 1577–1584
22. Tian, G.-W., Mohanty, A., Chary, S. N., Li, S., Paap, B., Drakakaki, G., Kopec, C. D., Li, J., Ehrhardt, D., Jackson, D., Rhee, S. Y., Raikhel, N. V., and Citovsky, V. (2004) High-Throughput Fluorescent Tagging of Full-Length Arabidopsis Gene Products in Planta. *Plant Physiol* **135**, 25–38
23. Sakamoto, K., and Briggs, W. R. (2002) Cellular and Subcellular Localization of Phototropin 1. *Plant Cell* **14**, 1723–1735
24. Marc, J., Granger, C. L., Brincat, J., Fisher, D. D., Kao, T., McCubbin, A. G., and Cyr, R. J. (1998) A GFP-MAP4 reporter gene for visualizing cortical microtubule rearrangements in living epidermal cells. *Plant Cell* **10**, 1927–1940
25. Wang, Y. S., Motes, C. M., Mohamalawari, D. R., and Blancaflor, E. B. (2004) Green fluorescent protein fusions to Arabidopsis fimbrin 1 for spatio-temporal imaging of F-actin dynamics in roots. *Cell Motil Cytoskeleton* **59**, 79–93

Index

MIX
Papier aus verantwortungsvollen Quellen
Paper from responsible sources
FSC® C105338

If you have any concerns about our products,
you can contact us on
ProductSafety@springernature.com

In case Publisher is established outside the EU,
the EU authorized representative is:
Springer Nature Customer Service Center GmbH
Europaplatz 3, 69115 Heidelberg, Germany

Printed by Libri Plureos GmbH
in Hamburg, Germany